Energy, Climate and the Env
Series Editor: **David Elliott**, E
University, UK

Titles include:

Manuela Achilles and Dana Elzey (*editors*)
ENVIRONMENTAL SUSTAINABILITY IN TRANSATLANTIC PERSPECTIVE
A Multidisciplinary Approach

Luca Anceschi and Jonathan Symons (*editors*)
ENERGY SECURITY IN THE ERA OF CLIMATE CHANGE
The Asia-Pacific Experience

Philip Andrews-Speed
THE GOVERNANCE OF ENERGY IN CHINA
Implications for Future Sustainability

Ian Bailey and Hugh Compston (*editors*)
FEELING THE HEAT
The Politics of Climate Policy in Rapidly Industrializing Countries

Gawdat Bahgat
ALTERNATIVE ENERGY IN THE MIDDLE EAST

Mehmet Efe Biresselioglu
EUROPEAN ENERGY SECURITY
Turkey's Future Role and Impact

Beth Edmondson and Stuart Levy
CLIMATE CHANGE AND ORDER
The End of Prosperity and Democracy

David Elliott (*editor*)
NUCLEAR OR NOT?
Does Nuclear Power Have a Place in a Sustainable Future?

David Elliott (*editor*)
SUSTAINABLE ENERGY
Opportunities and Limitations

Huong Ha and Tek Nath Dhakal (*editors*)
GOVERNANCE APPROACHES FOR MITIGATION AND ADAPTATION TO
CLIMATE CHANGE IN ASIA

Neil E. Harrison and John Mikler (*editors*)
CLIMATE INNOVATION
Liberal Capitalism and Climate Change

Horace Herring (*editor*)
LIVING IN A LOW-CARBON SOCIETY IN 2050

Matti Kojo and Tapio Litmanen (*editors*)
THE RENEWAL OF NUCLEAR POWER IN FINLAND

Antonio Marquina (*editor*)
GLOBAL WARMING AND CLIMATE CHANGE
Prospects and Policies in Asia and Europe

Catherine Mitchell, Jim Watson and Jessica Whiting (*editors*)
NEW CHALLENGES IN ENERGY SECURITY
The UK in a Multipolar World

Catherine Mitchell
THE POLITICAL ECONOMY OF SUSTAINABLE ENERGY

Espen Moe and Paul Midford (*editors*)
THE POLITICAL ECONOMY OF RENEWABLE ENERGY AND ENERGY SECURITY
Common Challenges and National Responses in Japan, China and Northern Europe

Marlyne Sahakian
KEEPING COOL IN SOUTHEAST ASIA
Energy Consumption and Urban Air-Conditioning

Benjamin K. Sovacool
ENERGY & ETHICS
Justice and the Global Energy Challenge

Joseph Szarka, Richard Cowell, Geraint Ellis, Peter A. Strachan and Charles Warren (*editors*)
LEARNING FROM WIND POWER
Governance, Societal and Policy Perspectives on Sustainable Energy

David Toke
ECOLOGICAL MODERNISATION AND RENEWABLE ENERGY

Thijs Van de Graaf
THE POLITICS AND INSTITUTIONS OF GLOBAL ENERGY GOVERNANCE

Xu Yi-chong (*editor*)
NUCLEAR ENERGY DEVELOPMENT IN ASIA
Problems and Prospects

Xu Yi-chong
THE POLITICS OF NUCLEAR ENERGY IN CHINA

Energy, Climate and the Environment
Series Standing Order ISBN 978-0-230-00800-7 (hb) 978-0-230-22150-5 (pb)

You can receive future titles in this series as they are published by placing a standing order. Please contact your bookseller or, in case of difficulty, write to us at the address below with your name and address, the title of the series and the ISBNs quoted above.

Customer Services Department, Macmillan Distribution Ltd, Houndmills, Basingstoke, Hampshire RG21 6XS, England

Climate Innovation
Liberal Capitalism and Climate Change

Edited by

Neil E. Harrison
Executive Director, The Sustainable Development Institute, USA

and

John Mikler
Senior Lecturer, University of Sydney, Australia

Editorial matter, selection, introduction and conclusion © Neil E. Harrison and John Mikler 2014

Remaining chapters © Respective authors 2014

All rights reserved. No reproduction, copy or transmission of this publication may be made without written permission.

No portion of this publication may be reproduced, copied or transmitted save with written permission or in accordance with the provisions of the Copyright, Designs and Patents Act 1988, or under the terms of any licence permitting limited copying issued by the Copyright Licensing Agency, Saffron House, 6–10 Kirby Street, London EC1N 8TS.

Any person who does any unauthorized act in relation to this publication may be liable to criminal prosecution and civil claims for damages.

The authors have asserted their rights to be identified as the authors of this work in accordance with the Copyright, Designs and Patents Act 1988.

First published 2014 by
PALGRAVE MACMILLAN

Palgrave Macmillan in the UK is an imprint of Macmillan Publishers Limited, registered in England, company number 785998, of Houndmills, Basingstoke, Hampshire RG21 6XS.

Palgrave Macmillan in the US is a division of St Martin's Press LLC, 175 Fifth Avenue, New York, NY 10010.

Palgrave Macmillan is the global academic imprint of the above companies and has companies and representatives throughout the world.

Palgrave® and Macmillan® are registered trademarks in the United States, the United Kingdom, Europe and other countries

ISBN: 978-1-137-31987-6 hardback
ISBN: 978-1-137-31988-3 paperback

This book is printed on paper suitable for recycling and made from fully managed and sustained forest sources. Logging, pulping and manufacturing processes are expected to conform to the environmental regulations of the country of origin.

A catalogue record for this book is available from the British Library.

A catalog record for this book is available from the Library of Congress.

To our families, for their love and generosity

Contents

List of Figures and Tables		ix
List of Abbreviations		xi
Acknowledgments		xiv
Notes on Contributors		xvi
1	An Introduction to Climate Innovation *Neil E. Harrison and John Mikler*	1
2	Institutions that Influence Climate Innovation *Neil E. Harrison and John Mikler*	19
Part I	**The National Context of Climate Innovation**	**41**
3	Climate Policy, Energy Technologies, and the American Developmental State *Robert MacNeil*	45
4	Colorado's New Energy Economy: Ecological Modernization, American-Style? *Stratis Giannakouros and Dimitris Stevis*	72
Part II	**The Corporate Context of Climate Innovation**	**97**
5	The Role of Corporate Scientists and Institutional Context: Corporate Responses to Climate Change in the Automobile Industry *David Levy and Sandra Rothenberg*	101
6	Corporate Investment in Climate Innovation *Neil E. Harrison and John Mikler*	127
7	US Labour Unions and Climate Change: Technological Innovations and Institutional Influences *Dimitris Stevis*	164
Part III	**Climate Innovation Across Borders**	**189**
8	The Influence of US Neoliberalism on International Climate Change Policy *Jeffrey McGee*	193

9	Varieties of Capitalism and US versus Chinese Corporations' Climate Change Strategies *John Mikler and Hinrich Voss*	215
10	Institutional Complexity in European Union Climate Innovation: European and National Experiences with Off-Shore Renewable Energy *Ian Bailey*	235
11	Conclusion: A Way Forward *Neil E. Harrison and John Mikler*	255

Bibliography — 275

Index — 305

List of Figures and Tables

Figures

Figure 1.1	The market innovation process	7
Figure 1.2	The climate innovation process	9
Figure 5.1	Articles mentioning climate change and the automobile industry, 1986–1999	108
Figure 8.1	Climate change mitigation policy spectrum	197

Tables

Table 2.1	Patents in climate change and energy technologies – Top fifteen OECD countries, 2003–2008	22
Table 2.2	Relative intensity of national policies and measures for renewable energy, 1999–2012	24
Table 2.3	Hypothesized national measures of climate innovation	35
Table 3.1	Energy Policy Act of 1992 alternative energy programs	50
Table 3.2	NCCTI targeting apparatus	53
Table 3.3	Selected developmental activities within individual agencies	54
Table 3.4	CCTP research and networking groups	58
Table 3.5	Common barriers to novel energy technology deployment	61
Table 3.6	Failed major climate bills in the US Congress, 2003–2010	64
Table 3.7	Selected failed non-comprehensive climate bills in the US Congress, 1999–2010	66
Table 5.1	Ford versus GM's interface with climate science, 1970s to late 1990s	116
Table 6.1	Governance	132
Table 6.2	Perceptions of climate change risks	134
Table 6.3	Perceptions of climate change opportunities	135
Table 6.4	Firms including climate change in risk assessments or strategy	138
Table 6.5	Percentage of targets by type	139

Table 6.6	Number of firms declaring annual absolute and intensity targets	140
Table 6.7	Company reports reviewed by CDP industry sector	160
Table 7.1	A typology of climate innovations	166
Table 7.2	Institutional factors influencing union preferences	168
Table 9.1	GHG emissions and global positioning	223
Table 9.2	MNCs' CSR and sustainability reports	225
Table 9.3	Chinese reports	228
Table 9.4	US reports	229
Table 10.1	National renewable energy targets for 2020, by member state	240

List of Abbreviations

AAMA	American Automobile Manufacturers Association
AFL-CIO	American Federation of Labor – Congress of Industrial
APP	Asia-Pacific Partnership
ARPA-CC	Advanced Research Project Agency – Climate Change
ARPA-E	Advanced Research Project Agency – Energy
ARRA	American Recovery and Reinvestment Act
ASEAN	Association of South East Asian Nations
BCT	Building and Construction Trades
BGA	BlueGreen Alliance
CACJ	Clean Air Jobs Act
CAFE	Corporate Average Fuel Economy
CCTP	Climate Change Technology Program
CDM	clean development mechanism
CDP	Carbon Disclosure Project
CEO	Chief Executive Officer
CFC	chlorofluorocarbon
CfDs	Contracts for Difference
CIBUL	Centre for International Business University of Leeds
CME	coordinated market economy
CNOOC	China National Offshore Oil Company
CO_2	carbon dioxide
CO_2e	carbon dioxide equivalent
COGCC	Colorado Oil and Gas Conservation Commission
COP3	Third Conference of Parties
COSEIA	Colorado Solar Energy Association
CPLI	Carbon Performance Leadership Index
CRADA	Cooperative Research and Development Agreements
CREC	Colorado Renewable Energy Collaborative
CRES	Colorado Renewable Energy Society
CSR	corporate social responsibility
CWA	Communications Workers of America
DARPA	Defense Advanced Research Projects Agency
DBS	Developmental Bureaucratic States
DJSI	Dow Jones Sustainability Index
DJSWI	Dow Jones Sustainability World Index
DNS	development network state

DOD	Department of Defense
DOE	Department of Energy
EC	European Commission
EERE	Energy Efficiency and Renewable Energy
EIR	Entrepreneur in Residence
EISA	Energy Independence and Security Act
EM	ecological modernization
ETO	Environmental Technology Opportunities Program
ETS	Exchange Trading Systems
EU	European Union
FiT	feed-in-tariff
GCC	Global Climate Coalition
GCRD	Global Change Research and Development Program
GDP	Gross Domestic Product
GEO	Governor's Energy Office
GHG	greenhouse gases
GM	General Motors
GSI	Geographical Spread Index
ICTF	International Clean Technology Fund
IE	industrial ecology
INSTAAR	Institute of Arctic and Alpine Research
IPCC	International Panel on Climate Change
IRR	internal rate of return
ISA	International Studies Association
IT	information technology
JI	joint implementation
JILA	Joint Institute for Laboratory Astrophysics
LIUNA	The Laborers' Union
LME	liberal market economy
LPO	Loan Programs Office
MEP	Major Emitters and Energy Consumers Process
MERiFIC	Marine Energy in Far Peripheral and Island Communities
MNCs	multinational corporations
Mte	metric tonnes equivalent
NCAR	National Center for Atmospheric Research
NCCTI	National Climate Change Technology Initiative
NEE	New Energy Economy
NREL	National Renewable Energy Lab
OECD	Organisation for Economic Co-operation and Development
OEDIT	Office of Economic Development and International Trade
ORD	Office of Research and Development

PIC	Policy and Implementation Committee
R&D	research and development
RDA	Regional Development Agencies
REAs	rural electric authorities
RES	Renewable Energy Standard
RMI	Rocky Mountain Institute
RO	Renewable Obligation
ROCs	Renewable Obligation Certificates
SASAC	State-owned Assets Supervision and Administration Commission
SBIR	Small Business Innovation Research
SEIA	Solar Energy Industries Association
SERI	Solar Energy Research Institute
SOWFIA	Streamlining and Ocean Wave Farm Impacts Assessment
STS	science, technology and society
STTR	Small Business Technology Transfer
TNI	Transnationality Index
UNCTAD	United Nations Conference on Trade and Development
UNFCCC	United Nations Framework Convention on Climate Change
US	United States
USCAP	United States Climate Action Partnership
USW	United Steelworkers
VC	venture capital
VCs	venture capitalists
VOC	Varieties of Capitalism

Acknowledgments

The genesis of this book was in discussions we had in 2008 while attending the International Studies Association's (ISA) 49th Annual Convention in San Francisco. John had been researching how the comparative capitalism literature, particularly that pertaining to the Varieties of Capitalism Approach, might be applicable to the question of climate change, and his book *Greening the Car Industry: Varieties of Capitalism and Climate Change* (2009) was nearing completion. Neil had a longstanding interest in technological innovation and alternative approaches to addressing environmental concerns beyond the mainstream economic orthodoxy, beginning with his book *Constructing Sustainable Development* (2000) and evolving through other publications. After several hours of stimulating discussion around our overlapping interests one evening, as well as the consumption of a very good Italian meal, we agreed to 'do something about capitalism, technological innovation and climate change'. Five years and many more conversations, conference papers and two journal articles later, this book is the result.

The book would not have been possible without the support of the ISA, which funded a Venture Workshop at its 53rd Annual Convention in San Francisco in 2012. Most of the contributors to this volume also attended this workshop and generated lively debate about the problem of climate innovation. We would also like to thank Mark Huberty, whose alternative perspectives contributed greatly to the evolution of the book's theme, but who was ultimately unable to contribute a chapter to the book. Our thanks also go to Matthew Paterson, Mark Beeson, Matt McDonald, Peter Christoff, Kate Crowley, Lorraine Elliott, Robyn Eckersley, Willem van Rensburg, Bruce Tranter, Simon Niemeyer, David Schlosberg and Stephen Wilks for their support, comments and suggestions, as well as to the anonymous reviewers of our publishing proposal and the penultimate draft of the book.

The contributors to this book are no doubt grateful for the support of many people who assisted them in undertaking their research. We personally would like to thank Maia Kutner for allowing us more than usual access to the Carbon Disclosure Project reports. For their assistance in undertaking interviews, we thank Tennant Read from the Australian Industry Group, and Cathy Bray. The time and fascinating insights offered by all the anonymous interviewees were also much appreciated. We thank Christina Brian and Amanda McGrath from Palgrave Macmillan for their support, understanding and advice throughout the entire process. Our long-suffering partners and families as always deserve our thanks for putting up with us while we completed this project, often Skyping at all sorts of strange hours of the day and night. Our thanks in particular go to Kara, Annika, Erin, and Ursula.

This has been a wonderful collaborative research experience. We enjoyed working together, as well as with all the contributors to the book to whom greatest thanks goes. Despite being on opposite sides of the world we have become good friends in the process. We therefore hope that this book is one of our first, rather than last words, on the topic, as well as stimulating the research agendas and contributing to the understanding of its readers.

Neil E. Harrison and John Mikler

Notes on Contributors

Ian Bailey is an Associate Professor at Plymouth University, UK. He has published widely on aspects of European, UK and Australian climate politics, carbon markets, marine renewable energy and the geographies of the green economy. His recent books (both with Hugh Compston) include *Feeling the Heat: The Politics of Climate Policy in Rapidly Industrializing Countries* (2012) and *Climate Clever: How Governments Can Tackle Climate Change (and Still Win Elections)* (2012). He has advised the UK government, World Bank, European Commission and Policy Network on various aspects of climate and energy policy.

Stratis Giannakouros is a graduate student at Colorado State University, US. His research interests are in renewable energy transitions at the national and sub-national level and their role in addressing climate change.

Neil Harrison is Executive Director of The Sustainable Development Institute, a privately funded research institute and think tank which he helped to found. His long-term research agenda is concerned with how complex capitalist societies may develop sustainably. In addition to many articles and book chapters he has published four books. His latest *Sustainable Capitalism and the Pursuit of Well-Being* to be published in 2014 proposes a radically new approach to sustainable development. His next research project focuses on how corporations can use an understanding of complex systems to profitably participate in sustainable development.

David Levy is Chair of the Department of Management and Marketing at the University of Massachusetts, US. He founded and is currently Director of the Center for Sustainable Enterprise and Regional Competitiveness, which engages in research, education and outreach to promote a transition to a clean, sustainable, and prosperous economy. His research examines corporate strategic responses to climate change and the growth of the clean energy business sector. More broadly, his work explores strategic contestation over the governance of controversial issues engaging business, states, and NGOs.

Robert MacNeil is a Lecturer at the University of Sydney, Australia, where he teaches environmental politics. His work on the developmental state and US climate policy has been published in *Environmental Politics* and *New Political Economy*, and is the subject of his current book project entitled *Neoliberal Climate Policy: From Market Fetishism to the Developmental State*.

Jeffrey McGee is a Senior Lecturer at the Newcastle Law School, Australia, and a Research Fellow of the Earth System Governance Project. His main research area is international environmental governance with particular focus on interaction between the United Nations climate regime and regional climate change institutions. He also teaches environmental law, legal theory and public international law at Newcastle Law School.

John Mikler is a Senior Lecturer at the University of Sydney, Australia. His research interests are primarily on the role of transnational economic actors, particularly multinational corporations, and the interaction between them and states, international organizations, and civil society. He is the author of *Greening the Car Industry: Varieties of Capitalism and Climate Change* (2009) and editor of *The Handbook of Global Companies* (2013). He has also published over twenty journal articles and book chapters, including (with Neil Harrison in 2012) 'Varieties of Capitalism and Technological Innovation for Climate Change Mitigation' in *New Political Economy*.

Sandra Rothenberg is a Professor at Rochester Institute of Technology's Philip E. Saunders College of Business in the Department of Management. She is the Director of the Institute for Business Ethics and Social Responsibility, the mission of which is to support research and teaching at Saunders that focuses on the interdependent relationship between business and society. Her research primarily focuses on corporate environmental strategy and management, with a focus on the auto and printing industries.

Dimitris Stevis is a Professor of world politics at Colorado State University. His long term research and teaching have been on transnational political economy and social governance with a focus on labor and the environment. His current research includes Global Framework Agreements between multinational corporations and labor unions, the environmental politics of labor unions, with particular emphases on equity and strategies for greening production, and Colorado's efforts to move to renewable energy, with particular attention to local-global linkages.

Hinrich Voss is a Roberts Academic Research Fellow at the Centre for International Business University of Leeds (CIBUL), UK. His research interests concentrate on international business strategies of multinational enterprises from developed and developing countries. He investigates how these firms are affected by climate change policies and institutional objectives and if and how they adapt their international business configuration. He is also conducting research on the internationalization and the international competitiveness of mainland Chinese companies.

1
An Introduction to Climate Innovation

Neil E. Harrison and John Mikler

The international negotiations for reducing greenhouse gas (GHG) emissions conducted under the purview of the United Nations have garnered much attention, both in the press and in the scholarly literature. However, the results of these negotiations have delivered little and global GHG emissions continue to rise. After Copenhagen, it became clear to all but the most optimistic that an effective global political agreement to mitigate climate change is very unlikely. If the past is a guide to the future, it seems more likely than not that future international negotiations will also fail. This point is made by authors such as Giddens (2011: 208), who notes that hopes for an effective international agreement rest largely on 'an illusory world community'. The efforts of the handful of nations that are the key GHG emitters are of particular importance, because whether or not negotiations succeed and an international agreement emerges, it will be their national governments that face the challenge of implementing policies to meet GHG emission reduction targets.

There is some cause for optimism in this respect. In the wake of the United Nations Framework Convention on Climate Change (UNFCCC), Annex 1 countries agreed to bear the brunt of the huge GHG emission reductions required to prevent a dangerous change in the global climate, and most national governments have established ambitious targets.[1] This is true even of the United States (US), which has failed to ratify the Kyoto Protocol but has recently proposed a non-binding target to reduce its emissions by 83 percent by 2050 (UNFCCC, 2011; Hassol, 2011). That is an essential but easy first step. What remains is to develop the means to meet those targets. Political constraints ensure that all governments – especially democratically elected ones – cannot significantly increase the economic costs of essential goods and services or sufficiently regulate social activities to meet their self-imposed or environmentally desirable GHG emissions reductions targets. Their preferred option will be to develop technological innovations to reduce GHG emissions – that is, technological innovation to mitigate climate change, what we call *climate innovation*.[2] The importance of climate innovation to combat climate change is not new. From the earliest days of the international climate change negotiations, there has been a general acceptance that technological innovation will be crucial to mitigating climate change. It therefore seems increasingly inevitable that climate innovation will have to emerge in distinct national contexts. The attraction of the technological innovation option is that it supports government efforts to maintain economic growth. But what is the potential for climate innovation in countries that embrace the most liberal conceptions of capitalism such as the US? This is the puzzle we investigate in this book.

Climate innovation and the market

Technological innovation is generally believed to be a primary strength of capitalism. In a reflection of the prevailing *zeitgeist* and the mandate of the 1997 Kyoto Protocol to the UNFCCC, studies – for example, those of the IPCC, inter-governmental agencies and private consultancies – assume that the market will provide the necessary solutions.[3] Rather than a comparative analysis that focuses on distinct national contexts, research into such questions has tended to produce generalizations for public policymakers that are increasingly being followed as received wisdom – for example, taxing emissions, subsidizing the uptake of new low-emitting technologies, introducing market based systems to trade emission rights, and so on. It is almost as if a 'Washington Consensus' for mitigating climate change has emerged, though in this case not from Washington. It is commonly assumed that if prices are made 'right' – that is, they include the cost of all externalities – all will be well. Therefore, it is now generally accepted by both economists and most governments that GHG emissions can be most efficiently mitigated – and technological innovations will spring forth that permit this – through market mechanisms. This is also true of much of the scholarly literature. For example, Newell (2008: 2) proposes a strategy for the US to generate technological innovations that would reduce GHG emissions based on the 'simple principle that, within a market-based economy, success is maximized if policies directly address specific market problems'. His prescriptions are for government to ensure a stable long-term carbon price and fund targeted climate mitigation (primarily basic) research.

As such, the conventional wisdom is based on mainstream economic approaches that treat GHGs as an environmental externality of production to be internalized via market mechanisms. Altering market signals should ensure that environmental costs become economic ones, and in theory various positive and negative incentives, should permit governments to pressure consumers to choose GHG mitigating products. However, there are many reasons why such prescriptions may be valid in theory but not in practice. Because they are fundamental to the functioning of advanced economies, activities like transportation and energy production that contribute most to greenhouse gas emissions are price inelastic and not susceptible to market mechanisms. The deeply embedded nature of emissions in all aspects of economic activity underpinning advanced industrialized nations is widely recognized, yet the implications of this seem under-acknowledged in the embrace of a liberal, market perspective of the problem of climate change and mitigating its impact.

For example, the energy sector accounts for around 83 percent of greenhouse gas emissions (UNFCCC, no date),[4] and a key component of the energy sector is transportation which on its own accounts for around 25 percent of total carbon dioxide (CO_2) emissions. Up to 85 percent of this is accounted for by road transport (UNEP, 2003; Paterson, 2000). Given that around 75 percent of CO_2 emissions over the lifecycle of any vehicle occur in use (Deutsche Bank, 2004: 58)[5] taxing fuel to alter price signals may be a key policy for CO_2 emissions of vehicles in use prior to encouraging the uptake of more efficient vehicles but would raise costs throughout the economy and may especially impact the poor. Even supply shocks such as the one in the latter half of the last decade when oil prices rose 300 percent between 2003 and 2008 (IEA, 2009) have had little impact on fossil fuel consumption. Indeed, a study by Small and Van Dender (2008: 182) concludes that 'short-run supply shocks have bigger price effects and that long-run demand will not be curbed strongly as prices rise'. Other studies such as Graham and Glaister (2002, 2004) estimate that UK fuel prices would have to rise more than incomes to affect fuel purchasing decisions, and that even if one holds income constant a substantial 10 percent rise in the price of fuel produces only a 3 percent fall in fuel consumption. In the US around where 90 percent of travel is by motor vehicle, higher fuel prices only serve to increase costs to consumers who have little choice but to rely on their automobiles (OECD, 1996; Harrington and McConnell, 2003).

Data also suggest that even if this were not the case and policies to drive price signals for viable low carbon technology alternatives were more politically tractable, these may still have little effect for reasons beyond political and economic considerations. For example, in the EU the price differential between gasoline and diesel fuel is often seen as a key reason for the uptake of more efficient advanced diesel vehicles over the past decade to the extent that they now account for half of all new vehicle purchases. However, a snapshot of the gasoline-diesel price differential in EU member states versus the share of diesel-powered vehicles demonstrates that there is little relationship between the two. For example, gasoline is over 50 percent more expensive than diesel in Austria, Belgium, Finland and the Netherlands, yet the share of diesels in total car sales in these countries varies substantially from 20 to 62 percent. Furthermore, although the percentage gasoline is more expensive than diesel fell from 44 to 37 percent on average across all EU countries over 2000–2006, the share of diesels in new car sales rose from 33 to 51 percent (OECD, 2007; ACEA, no date). In Japan there is a price differential of 46 percent, similar to that found in the EU, and it may be noted that this is a long standing feature of the Japanese market (OECD, 2007),[6] yet virtually no diesel vehicles are sold there (JAMA, 2011).[7] In the case of the US, even though diesel is marginally more expensive than gasoline, there would be little financial penalty in switching to diesel vehicles in terms of filling up a tank, while the cost benefits of switching to more fuel efficient diesel vehicles should have provided incentives to do so. As with Japan, this has long been the case but, as with Japan, it has not (United States Environment Protection Agency, 2010).[8]

There may be a range of reasons for consumers being unwilling to act 'rationally' on the basis of market price signals, such as US and Japanese consumers perceiving diesel vehicles as undesirable by comparison to gasoline powered vehicles with which they are more accustomed, while European consumers have long been more familiar with them and more inclined to consider purchasing them, regardless of sub-regional variations. But putting it simply, if consumers' behavior reveals their underlying preferences, consumers in some European states are more inclined to buy diesels than others, while those in Japan and the US simply do not want to consider diesels regardless of the economic benefits. Putting it more technically, the normative basis for preferences, many of which are socially constructed, means that these can become institutionalized over time.

With such fundamental sectors as transport characterized by high price inelasticity of demand, the market mechanisms necessary to drive technological innovation for climate change are therefore either so politically challenging (for example, they involve very high taxation of emissions) or so politically infeasible (for example, they require high subsidies in a time of global economic challenges or direct regulation of consumer choice) as to make a reliance on them unwise. Climate change may originate from unrestrained capitalism, as Newell and Paterson (2010) argue, but a liberal preference for relying on the market and market mechanisms to fix the problem is dubious because it is not an artifact of market demand.

Market innovation versus climate innovation

Technological innovation in advanced industrial countries is the product of the iterative interaction of science possibilities with market demands, and is primarily the result of choices by profit-seeking firms. This model of technological innovation has been well accepted as appropriate and applied across a range of countries (for example, see Dosi, 1982; Freeman, 1992). In market economies, firms profit by meeting customers' demands better than their competitors. For example, Apple is a profitable innovator because it has designed simple solutions to market needs using cutting-edge technologies, while the Toyota Prius, initially designed to demonstrate a technological solution to energy consumption, has created a niche market for hybrid fossil fuel/electric transportation systems.

Occasionally, science develops new knowledge that can drive the development of a whole new range of technologies, a new techno-economic paradigm. For example, the transistor and integrated circuit made digitization of control systems possible and seeded the current Information Age. Beyond Apple iPhones, information management systems have profoundly changed production systems, marketing, and social relations. From information technology (IT) have sprung such varied systemic changes as the automation of custom manufacturing; increases in the energy efficiency of production systems; the Internet and World Wide Web; and simulations of building and automobile designs, nuclear explosions, and the effects of climate change. Whole new industries have emerged and old economy industrial companies have been replaced by entrepreneurial upstarts that earn billions from search engines and on-line advertising, a classic example of the 'creative destruction' of capitalism (Schumpeter, 1961[1942]).

The conventional market innovation model reflects innovation from both *demand-pull* and from *science-push*. The distinction is only a matter of emphasis. Demand-pull suggests that market signals cause a search for technological solutions and science-push would also be ineffective if the new technologies that science produces were not better able to meet market demands than current technologies. We use information technology because word processor software on a computer is more flexible than typewriters, because digital technology is more readily adaptable than the analog technologies that preceded it, and because computing systems can render the complex mathematical calculations for simulations infinitely more rapidly than abacuses.[9] In each case, the IT solution satisfies the primary consumer needs of document preparation and printing; cheap, flexible manufacturing; and system design and prediction better than the prior technology.

Markets and firms' responses to their economic environment may explain the total amount of innovation but they do not explain the process of innovation or the activities that that process encompasses. If a firm sees within its market a threat or an opportunity which an innovation can curtail or seize, how does it develop that innovation and use it in the market? Early models of technological innovation presumed that the dominant cause was the push from scientific discovery, driven by government financing or the lone inventor is his garden shed (Bernal, 1989; Rosenberg, 1994; Bush, 1945). As new scientific knowledge is produced, eventually some entrepreneur or entrepreneurial entity finds a way to put it to use in satisfying human needs. Later theories argued that market demand determined technological innovation (Schmookler, 1966; Mowery and Rosenberg, 1979). Innovation is now recognized as a complex, iterative process with multiple internal positive and negative feedback loops that connects developments in scientific knowledge with market demand (Freeman, 1979; Keller, 2008).

Figure 1.1 is a simple, generalized model of the 'normal' market-driven innovation process. Firms in most industrial sectors will follow much the same steps in a process that translates scientific knowledge into products and services that meet consumer needs. However, although some commentators such as Grubb (2005), a leading scholar of climate change, and to some extent Pielke Jr (2010) apply much the same model as shown in Figure 1.1 to the process of innovation to address climate change, we contend that the innovation process shown in Figure 1.1 is not directly applicable to climate innovation. This is because the process for delivering technologies for climate innovation

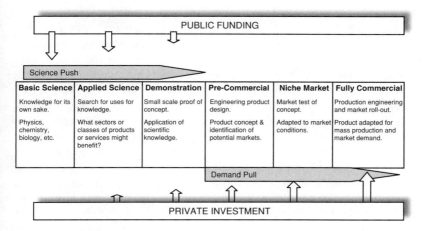

Figure 1.1 The market innovation process

is different from the 'normal' process in three important ways: it demands inter-disciplinary research; it must meet technical targets rather than market demands; and, relatedly, consumers do not demand its effect, GHG reductions.

First, regarding inter-disciplinary research, useful and necessary innovations may be found in every aspect of modern life from agriculture through construction to transportation and may involve technical knowledge and production techniques from several scientific disciplines and multiple industries. For example, reducing GHG emissions in transportation may involve research in materials, fossil fuel combustion, tire design, and chemistry. This means that the disciplinary silos in universities and industry specific research by firms are unlikely by themselves to be as successful as distributed innovation networks or custom constructed inter-disciplinary research facilities exemplified by Bell Labs in its heyday (Milford and Barker, 2008).

Second, to meet GHG emissions reductions targets, nations cannot rely on market demand. They must invest in the development and diffusion of a range of technologies that will reduce their emissions by a defined amount, about 80 percent by 2050 for the industrialized countries as a group (Hassol, 2011). Satisfying an environmental necessity is very different from satisfying the demands of human consumers. Consumers may demand goods and services that are cheaper, faster, smarter, and easier but the 'only' punishment for failure may be bankruptcy of a firm and loss of jobs. The environment is expected to be much less forgiving, if emissions reductions targets are not met.

Third, consumers do not demand technologies that are developed specifically to mitigate GHG emissions. Indeed it is unlikely that they ever will. They may demand technologies that improve their lives that also reduce their 'carbon footprint' but they do not expect to have to manage their footprint directly. They may select the most environmentally efficient products but do so as long as they satisfy other more important personal needs they have. This is suggested in surveys of social attitudes (for example, see Mikler, 2011),[10] as well as in analyses of product development. For example, a recent study by Knittel (2011) shows that nearly all the advances in vehicle engine technologies since 1980 that could have gone into reducing CO_2 emissions have gone instead towards heavier, more powerful vehicles. In fact, he notes that 'if weight, horsepower, and torque were held at their 1980 levels, fuel economy for both passenger cars and light trucks *could have* increased by nearly 60 percent from 1980 to 2006. This is in stark contrast to the 15 percent by which fuel economy actually increased' (Knittel, 2011: 3368–3369). The technological innovations that could have delivered low carbon mobility *by now*, have instead been largely directed by the car industry as a whole towards marketing opportunities, an example of innovation driven by competitive markets and consumer demand. If automobiles with the lowest emissions are less effective at meeting the transportation or image needs of customers, the firms cannot be blamed for failing to invest in new technologies for products that they cannot sell.

For these reasons the process of climate innovation looks more like that in Figure 1.2. Although demand-pull will be weak, firms cannot rely on consistent or permanent government interventions to enhance market demand for climate innovations. Therefore, subsidies and taxation or regulation of competing products can only reduce the risk of investing in the diffusion of current or incremental climate innovations with relatively short timeframes. Even then, these interventions in markets will not 'pull through' many climate innovations and will have no effect on investment decisions for more radical innovations. Governments are unable to guarantee market demand for any particular climate innovation short of contracting to purchase it in commercial quantities at some fixed, future time. Even then, future governments may renege, as purchases of products and services embodying climate innovations would only be effective in the short term for available technologies when government need is evident.

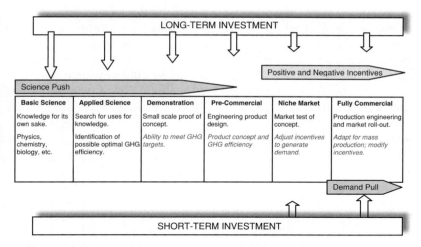

Figure 1.2 The climate innovation process

More radical climate innovations that require longer lead times and government interventions to spur market demand are rarely considered sufficiently permanent because their effect on demand is difficult to predict and political needs change. For example, the German Renewable Energy Sources Act came into force in 2000 and photovoltaic feed-in tariffs became effective from 2004, but their rates have changed several times since then to adjust the supply of renewable energy investment to the government's budget (Frondel et al., 2009). Although the Obama Administration has proposed to make the current temporary residential renewable energy tax credits permanent, a future Congress can readily change the credit or eliminate it.[11]

As such, the drivers for, and process of, climate innovation are different. Climate innovation demands technological innovations whose express purpose is to mitigate climate change. Only by developing technologies for the purpose of reducing GHG emissions by a targeted amount, in order to avoid dangerous climate change, can those targets be reached without governments mandating social and behavioral changes that are much less politically acceptable. Markets by themselves are unlikely to demand in sufficient quantity the climate mitigating technologies to sufficiently prevent dangerous climate change. To reach specific GHG emissions targets while demand-pull is weak firms will have to rely more heavily on science-push than in the conventional model. Consequently, they will need to invest more heavily in scientific and developmental research with more uncertain outcomes over a much longer period than is necessary for market products. Without clear market signals, what influences decision-makers in governments and private firms to invest in climate innovation? In practice, answering this question involves studying the activities not just of governments, but also of research laboratories and firms as they contribute their expertise to the innovation process and help to select among alternative technology paths. And it involves studying them in their distinct national contexts.

National climate innovation systems and US liberal capitalism

There is a limit to the extent to which we can delve into the personal psyches of decision-makers, but we can assess the network of institutions – the 'national innovation system' – that establishes the 'rules of the game' for technological innovation. Therefore, this book assesses the ability of a liberal capitalist national innovation system, such as that exemplified by the US case, to generate the radical and incremental climate mitigating technologies that the country will need to avoid social engineering while appropriately addressing the global challenge of climate change. In so doing, the book's contributors recognize that there are nationally specific social and cultural institutions that influence the behavior and choices of individual decision-makers within firms and the innovations in which they choose to invest. Thus, we assume that capitalism as it is practiced will continue to vary between countries because of the prior institutional structures that establish the range of possibilities for future evolution.

There is much discussion of 'national innovation systems' but most of it is founded on a distinctly economic perspective. For example, the OECD (1997: 9) comments that their underlying premise is that:

> innovation and technical progress are the result of a complex set of relationships among actors producing, distributing and applying various kinds of knowledge. The innovative performance of a country depends to a large extent on how these actors relate to each other as elements of a collective system of knowledge creation and use as well as the technologies they use. These actors are primarily private enterprises, universities and public research institutes and the people within them.

Others define the national innovation system in terms of the network of institutions in 'the public and private sectors whose activities and interactions initiate, import, modify and diffuse new technologies' or as 'that set of distinct institutions which jointly and individually contribute to the development and diffusion of new technologies and which provides the framework within which governments form and implement policies to influence the innovation process' (OECD, 1997: 10 from Freeman, 1987, 1995; Metcalfe, 1995). In every case the focus is then on the interactions among a network of *economic* actors, that may include state actors, that constitutes the system of innovation in a nation. The system is the accumulation of behaviors without consideration of why individual economic actors choose to behave as they do. This perception of the nature of the national innovation system developed from an interest in the economic effects of technological innovation. As such, it reflects a distinctly economic disciplinary approach that expects the supply of positive incentives (primarily from government) and the demand from the market to determine both the quantity and quality of technological innovation.

In Mikler and Harrison (2012), we noted that there is a surprising dearth of scholarship from a comparative political economy perspective. Yet, capitalism is not monolithic. The formal and informal institutional basis for capitalist relations of production varies considerably between states. Is capitalism as it is practiced in the largest GHG emitting countries up to the challenge of delivering the necessary technological innovations? Authors such as Newell and Paterson (2010) and Koch (2012) argue that without substantial changes it is not. In this book we open up the concept of capitalism to encompass dimensions beyond the economic that may help to design policies to generate technological innovations that will significantly mitigate climate change and rapidly diffuse them throughout society. Meadowcroft's (2005) comprehensive overview of scholarly developments in the environmental political economy literature demonstrates that approaches to technological change have focused on issues such as trade and the environment; the merits of government regulation in the form of direct regulation versus market mechanisms; the greening of production via business itself exemplified by such concepts as corporate social responsibility; the impact of the Kyoto Protocol; and social concern for sustainable consumption. The theoretical lenses applied to study such issues have had both normative and rationalist epistemological underpinnings – for example, ecological modernization versus solving collective action problems inherent in managing the commons (for example, see Mol et al., 2000; Mol and Sonnenfeld, 2000; Ostrom, 1990). Yet, the relevance of the vast comparative capitalism literature has remained largely unchartered waters. What is needed is an analysis of the nature and effect of the nationally defined institutional contexts in which profit-seeking firms decide to invest in climate innovation. With a primary focus on the United States, this book begins such an analysis.

In keeping with our goal to get 'under the hood' of capitalism and examine the processes of technological innovation directly, we understand the national innovation system as the network of institutions that forms the context within which firms and states choose to invest in particular technological opportunities. We are agnostic about the nature of the institutions that form the institutional context of innovators. Certainly, government subsidies for research and development (R&D) will play a part as will venture capital at some stage in the process. Eventually any product must find or create a market demand, and climate innovations will be no different, but demand for such products cannot be effectively anticipated so it must play a relatively insignificant role in the decision to invest in climate innovations. In our view the institutions that comprise the national innovation system may be social (for example, pressures from environmental activist groups), cultural (for example, the acceptance of, and reward to, entrepreneurs), and political and legal (for example, corporate governance and labor law). In short, we hypothesize that the institutional context is more complex than most economists can see. But most importantly, because the climate innovation process in Figure 1.2 is quite different from the 'market innovation' process in Figure 1.1, it is reasonable to suppose that the institutional arrangements of an effective national innovation system for climate innovation would be different from current systems that are predicated on substantial market demand, and that they will vary substantially between capitalist states.

Plan of the book

In Chapter 2, we develop a theoretical framework which categorizes the types of institutions within the national innovation system that enable and constrain investment in climate innovation. First, we show how innovation for climate change must differ from innovation in response to market needs. Then, we break the cluster of institutions that comprise the national innovation system into four 'dimensions' to propose a taxonomy of categorical qualities that would help or hinder climate innovation. For example, we argue that patient finance is needed to develop the radical climate technologies that will be needed to mitigate the GHG emissions that probably cannot be reduced by relying on incremental improvements on current technologies. But current venture capital or private equity funds that have been effective at developing new Internet enterprises like Google and Facebook do not have the patience to invest in long-term technological innovation of energy and production systems for substantial GHG emission reductions.

Following this, the chapters are arranged in thematic parts. At the beginning of each part, there is a short introduction that summarizes the principal findings of each chapter, shows how the chapters relate and integrate, and discusses possible tentative conclusions. The concluding chapter draws together the conclusions of each section into a coherent whole and develops recommendations from their interaction.

Part I is on the national context of climate innovation and has two chapters. In Chapter 3, Robert MacNeil examines the domestic US response to the climate challenge and investigates the developmental state apparatus being deployed in Washington's efforts to develop markets for novel climate and energy technologies. He uses a regulationist framework to understand how critical tensions and contradictions inherent in liberal market economies such as the US serve to engender new forms of direct state intervention, particularly within the innovation process. Building on Block's (2008b) concept of a 'hidden developmental network state' emerging in Washington beginning in the late 1970s, his chapter explores the tensions that have led to the use of such developmental policies in the climate and energy realms, and provides a description of how these policies function.

In Chapter 4, Stratis Giannakouros and Dimitris Stevis focus on a sub-national effort to create a renewable energy cluster in Colorado. They do so in order to examine the interplay of US federal and sub-national institutions in climate innovation. The US constitution reserves many powers to the state except to the extent that the requirements of national commerce require a federal response. While the federal government is often charged with overreaching to grab power from the states, the states retain significant flexibility in industrial and environmental policy. The case of the state of Colorado is used to highlight the contribution of sub-federal institutions to addressing environmental and economic challenges. First, they develop an analytical scheme within which to place policy initiatives on a continuum from grey industrial policy to strong ecological modernization policy by identifying key institutions that are influential in each policy type. This analytical scheme is then applied to the transitional renewable energy policy period between 2004–2012 in the state of Colorado. This period starts with the adoption of a renewable energy portfolio in 2004 and includes the 'new energy economy' period from 2007–2010 as well as the years since. Looking at three key turning points this paper interprets the 'new energy economy' strategy using the analytical scheme developed and identifies the political and social institutions that frame this transition. Drawing upon these findings, the paper analyses the implications of the Colorado case for understanding sub-federal initiatives in the US more broadly and concludes that such initiatives are insufficient to generate the climate innovation necessary given a national liberal economic context.

Part II considers the corporate context of climate innovation. In Chapter 5, David Levy and Sandra Rothenberg consider how internal structure influences how the challenge of climate change is understood and reported to executives. They argue that corporate strategies can be understood in terms of different institutional contexts, particularly the national political-economic setting, the competitive environment, each company's institutionalized historical narrative, and the organizational location of corporate scientists. They present the results of in depth qualitative research of the US automobile industry over the period when climate change first emerged as an issue of strategic relevance to corporate decision-makers. In particular, they focus on the key role played by corporate scientists who acted as 'boundary spanners' and institutional entrepreneurs, translating the wider climate science discourse and shaping corporate perceptions and strategic responses. As a result of interviews conducted, they find that corporate scientists who are more autonomous and better connected to the external climate science community are more likely to believe in the science and be advocates for emission-reducing innovation. Yet there is not a simple causal line from climate beliefs to action. Rothenberg and Levy identify a process of co-construction of climate science beliefs and economic interests and strategies, in that beliefs about climate change are themselves shaped by corporate (and departmental) perceptions of economic interest and strategic commitments to particular products.

In Chapter 6, we add to our understanding of corporate responses to institutional contexts through an analysis of US reporting for the Carbon Disclosure Project and interviews with senior office-holders from Australian corporations included in the Dow Jones Sustainability Index. Our analysis builds on that of Levy and Rothenberg's as we consider the current situation in which the science of climate change has been accepted by corporations, and as such they are now grappling with their strategic responses to the challenges it presents. Theory suggests that because these firms are based in liberal market economies their decisions to invest technological innovation of any kind will primarily be influenced by their perception of market demand. However, we find that they see market demand signals for climate innovation as weak or absent. Consequently, we demonstrate that companies investing in climate innovation, particularly in innovations that are radical and have a high science component and long lead time, must use different metrics for gauging the profitability of potential products and services. In particular, we show that this may necessitate more, and more extensive, government intervention but, more importantly, intervention that allows companies to better predict future regulatory requirements.

In the United States, private sector unions have been in decline for three decades and industrial relations are frequently combative. In short, the institutional environment is not supportive of unions or labor. In such an environment, it might be presumed that unions would narrowly concentrate on optimizing their membership and protecting jobs and oppose efforts to protect the natural environment. Yet, some unions have adopted long-term social and technical innovation programs responsive to different levels of climate innovation. In Chapter 7, Dimitris Stevis begins to explain these 'unexpected' results. Because unions are agents choosing their actions within an institutional context, many factors may influence their choices to support or reject climate innovation. These factors he analyzes in depth and with cogent examples to demonstrate that unions are most usefully theorized as purposeful and learning organizations in the same way that we would look at any other type of organization, and their learning is influenced by their institutional context but not determined by it. While this explains the variability of unions' responses to the political pressures they have faced in recent decades and the consequent effects on their membership and influence, it does not explain why any particular union would either support or reject climate innovation. Therefore, Stevis looks in depth at the unions that have helped to create the BlueGreen Alliance with leading environmental organizations and some industrial companies. By analyzing the actions of the Alliance and their publications, and interviewing officials, he is able to show that those unions made a political decision to attempt to effect change in the institutional context of labor relations through environmental action that includes climate innovation.

In a globalizing world no nation, however large and economically dominant, is an island unto itself. As globalization advances, what impact does crossing borders have on firms' climate innovation decisions and national policies? This is the focus of the contributions in Part III. Not only is the US widely considered an outlier even among liberal market economies, its influence in international negotiations has reflected a strong liberal bias in favor of market-based rather than political solutions. In Chapter 8, Jeffrey McGee argues that since the start of the new millennium the US has again changed its tactics to support technological solutions and reduce governmental coordination in markets for advanced mitigation technologies and practices. He shows how this technology-focused preference for climate governance reflects not a global, but a US neoliberal preference for economically viable solutions both domestically and internationally. Perhaps more importantly, through the lengthy international negotiations he demonstrates how the US discourse has been increasingly adopted by other countries including the more coordinated market economies of Europe.

The US and China are the two largest economies, the two largest emitter of GHGs, and are increasingly economically interdependent. Therefore, the strategic investment decisions of Chinese and American corporations play a central role in the extent and speed of climate innovation. In Chapter 9, John Mikler and Hinrich Voss combine perspectives from comparative political economy and international business strategy, supported by a content analysis of the reporting of leading US and Chinese corporations. In the process, they demonstrate that despite being multinational in their operations, these corporations remain institutionally embedded in their home states, and therefore their strategic decision-making on climate innovation is made through different 'lenses'. From their analysis, they suggest that MNCs' home state variety of capitalism and stage of development influences how these firms operate overseas, and what kind of organizational routines and practices they transfer abroad.

In Chapter 10, Ian Bailey uses the case of renewable energy to describe the institutional complexity that underlies European Union (EU) policymaking and strongly influences firms' innovation decisions. The principal EU policy tool for stimulating climate innovation and reducing emissions is its Emissions Trading Scheme, an idea first tested on a large scale in the US in 1990 amendments to the Clean Air Act that is wholly consistent with the liberal capitalist principles. His chapter then reviews attempts by the UK and French governments in particular to secure footholds in the ocean energy sector (offshore wind, wave and tidal) and, more generally, to restructure their energy sectors. In addition to highlighting interplays between European, national, regional and private-sector institutions during the construction, contestation and interpretation of the EU's climate and energy package, and the different methodologies used by member-state governments to decarbonize energy production, his chapter draws attention to the complex and uncertain processes through which technological innovation strives to achieve commercialization and navigate other components of the regulatory landscape.

In our concluding chapter, we begin by summarizing the results of the analyses in the foregoing chapters, and then relate these back to the theoretical framework we proposed in Chapter 2. We show that the US innovation system is not designed to support climate innovation. It is partially deficient in three of the four institutional 'dimensions' of climate innovation in our theoretical framework, and wholly deficient in the fourth. The US might be able to create institutions to ensure better long-term funding of applied science, bend the autonomy of corporate managers to expand their goals beyond a narrow liberal market concern for profits and investor returns, and reduce regulatory uncertainty. We do not expect that the US will achieve sufficient unity of purpose to achieve an effective collective response to climate change. However, we believe that even without this 'climate collectivism' the US should be able to construct the institutions necessary to substantially increase its climate innovation. Nevertheless, we are forced to conclude that liberal capitalist economies cannot be the most effective generators of climate innovation; that the European Union, though comprised of many non-liberal capitalist states, is prevented by subsidiarity from being significant more effective; and that the rising development states of Asia may be better placed to create and market the technological innovations that the world needs to avoid dangerous climate change.

While each country has evolved its own unique institutional framework in support of technological innovation, we expect that many of the lessons from the examination in this book of the US innovation system would apply. However, this is yet to be confirmed by further research. Thus, we develop an agenda for a comparative institutionalist research program across multiple leading national GHG emitters to identify the critical institutions in each nation that may need to be modified in order to stimulate climate innovation.

Notes

1 The forty-two Annex 1 countries include all members of the Organisation for Economic Co-operation and Development (OECD) as of 1992 (when the UNFCCC was signed) and economies then 'in transition' from Soviet central planning. See UNFCCC (2011) which lists self-imposed national targets stated at the 34th Session of the Subsidiary Body for Scientific and Technological Advice of the Conference of the Parties to the UNFCCC.
2 Polls of US public opinion vary depending on the questions and their context. On balance is seems that a slight majority of the US public believes that climate change is happening but fewer think it is an important problem threatening their future or that it is the 'most important problem facing the country today'. See polls reported at http://www.gallup.com/tag/Climate+Change.aspx and Dye (2010).
3 For example, in several studies McKinsey and Company, the management consulting firm, explicitly argues for only market-based systems, while such a perspective underpins much of the analysis of the OECD (for example, OECD, 2009, 2012).
4 This is based on 2011 data. The other main sectors are industrial processes (7 percent); agriculture (8 percent); and waste (3 percent).
5 This dwarfs CO_2 emissions at all other stages. For example, the manufacturing process accounts for only 2 percent of a vehicle's total lifecycle CO_2 emissions.
6 Gasoline fuel has been between 43 and 69 percent more expensive than diesel between 1980 and 2006.
7 The share of diesels in new vehicle purchases in Japan was 0.05 percent in 2006.
8 The share of diesels in new vehicle purchases in the US was 0.1 percent in 2008.
9 The effect of digital automation on labor and the dignity of work was foreseen in Vonnegut (1952), and critiqued in Shaiken (1986).
10 In this article, World Values Survey data are presented which show respondents are often unwilling to pay more for products that are more environmentally friendly, and are much more willing (up to twice as much in the case of the US and Japan) to choose products that are better for the environment but only if they are the same price.
11 According to the summary of the tax proposals of the 2012 presidential candidates available at http://www.cpa2biz.com/Content/media/PRODUCER_CONTENT/Newsletters/Articles_2012/CorpTax/CorporateTaxObamaRomney.jsp, President Obama's presumed challenger Mitt Romney does not support renewable energy tax credits, nor do Congressional Republicans.

2
Institutions that Influence Climate Innovation

Neil E. Harrison and John Mikler

Because of continuing political opposition to effective climate policy, the United States (US) is committed to mitigating climate change primarily through technological innovation. The challenge it now faces is how to stimulate innovation of those technologies that would be most effective at reducing its GHG emissions without reducing economic growth or personal freedom. Technological innovation is generally accepted as a primary constituent of economic growth (Romer, 1986; Solow, 1957) and the Obama Administration used nearly 10 percent (or more than $60 billion) of the funds from the American Recovery and Reinvestment Act of 2009 on various initiatives to stimulate green energy production.[1] Not only is technological innovation considered the primary tool to meet the US's non-binding target to reduce its GHG emissions by 83 percent from 2005 levels by 2050, but it also is seen as a solution to potential economic stagnation (Mikler and Harrison, 2012). This highlights the puzzle we are concerned with here: how do institutions support or undermine the climate innovation that is necessary in liberal capitalist economies? In this chapter we build a theoretical framework that categorizes the institutions that potentially impact the required technology innovation decisions.

As outlined in Chapter 1, we presume that technological innovations are generated in a firm-centered process that functions within a network of institutions. We further suggest that the technological innovations needed to mitigate climate change will require a different innovation process than the innovation process followed today for products and services designed to profitably satisfy market demand. Normally, firms innovate to meet consumer preferences for a better product or service, or an equivalent product or service at a lower price, than competitors. In other words, the exigencies of the market force perpetual innovation in both products and internal processes (Nelson and Winter, 1982). But this is only true up to a point. The quantity and quality of firms' innovations is also influenced by non-economic factors including the political institutions that govern the structure and operation of firms and the social institutions that influence how executives make decisions and how they handle uncertainty and risk. Because market demand usually is less important in decisions to invest in climate innovation, we argue that these non-economic institutions may have greater influence on firms' decisions. Therefore, our theoretical framework presumes that climate innovations are generated in a process that functions within a network of economic, political and social institutions. Climate innovations require a different innovation process from the innovation process for products and services designed to profitably satisfy market demand. This then supposes that the network of institutions within which firms decide to innovate will have to change to generate climate innovations sufficient to meet GHG emissions reductions targets rather than to satisfy short-term market demands.

In this chapter, we build our theoretical framework of the institutional context of climate innovation by starting from a perspective that although mainstream economic approaches treat GHGs as an environmental externality of production to be internalized via market mechanisms, in practice there are practical problems with such a viewpoint. Instead, we consider the role of institutions in technological innovation and make the case for why these should be central to an analysis of climate innovation. We then consider those types of national institutions that are most likely to impact the innovation process for climate mitigation in the developed states that caused the problem, and which have the technical and financial capacities to tackle it. Based on these analyses we conclude by presenting a four dimensional theoretical framework that explicitly acknowledges the manner in which the innovation process may vary between industrialized states and in response to the needs of national emissions targets and climate innovations.

National institutional variations and climate innovation

Because of its obsession with equilibrium, conventional economics paints a static picture of the economy that cannot account for the effects of commercialization of radical products (David, 1985; Arthur, 1989). Its obsession with a narrow conception of markets leads it to overlook the varieties of market and non-market arrangements that channel economic activity (Nelson, 2003). While economic theories have increasingly recognized the role of innovation in growth, they treat it as peripheral to economic activity, as developed outside the economy. Simplistic economic models do not capture the importance of design and entrepreneurial talent in connecting demand to knowledge (Walsh, 1984). Most economic theory does not look inside the firm to the many forces that influence when and how they innovate (Rosenberg, 1982). Finally, economic theory does not look past price to quality and other product characteristics that consumers may prefer (Pavitt, 1980).

Nationally specific economic, social and cultural institutions influence the behavior and choices of individual decision-makers within firms and the innovations in which they choose to invest. Thus, we assume that capitalism as it is practiced will continue to vary between countries because of the prior institutional structures that establish the range of possibilities for future evolution. In Mikler and Harrison (2012), we demonstrated that a strong focus on a purely market case, or market mechanisms, in a liberal market economy (LME) setting is likely to be less effective in encouraging the necessary radical technological innovations.[2] We found that although the US leads in research and development (R&D) because of the role of the military, and because it remains the largest high-income country in the world, it is also the case that coordinated market economies (CMEs) of Europe and East Asia are more research-intensive.[3] They allocate a larger share of GDP to research than the US, and Japan has proportionately more researchers. Also, by comparison to the CMEs, the other LME countries like the UK, Canada, Australia and New Zealand fare poorly in overall research effort based on percentage of GDP measures and (in some cases) the number of researchers as a percentage of the population. The result is not that LMEs produce fewer technologies than CMEs for mitigating climate change. The OECD data on patents in key climate change and energy technologies presented in Table 2.1, demonstrate that both are very capable of doing so, as there are both LMEs and CMEs in the top fifteen OECD countries based on patents in key climate change and energy technologies as a percentage of total patent applications. However, the CMEs in the top fifteen generally have a higher share of their patents for climate change and energy technologies than the LMEs. And it is perhaps astonishing to note in particular that regardless of the number of patents it may have filed, the US is not represented in the top fifteen countries for patents in

Table 2.1 Patents in climate change and energy technologies – Top fifteen OECD countries, 2003–2008

	Electric and Hybrid Vehicles		Energy Efficiency in Buildings and Lighting		Renewable Energy	
	Rank and % of total patents	No.	Rank and % of total patents	No.	Rank and % of total patents	No.
LMEs	(6)Canada, 0.35	16,153	(5)Canada, 2.0	16,153	(7) Ireland, 1.5	2,072
	(13)Australia, 0.20	11,926			(9) Australia, 1.5	11,926
	(14)US, 0.15	281,820				
	(15)UK, 0.15	36,459				
CMEs	(1)Japan, 0.80	151,193	(1)Netherlands, 4.0	20,241	(1) Denmark, 4.5	7,052
	(2)Germany, 0.60	100,247	(3)Japan, 2.5	151,193	(6) Norway, 3.0	3,690
	(9)Austria, 0.40	6,950	(7)Austria, 1.5	6,950		
	(10)Sweden, 0.30	15,780	(8)Germany, 1.5	100,247		
			(13)Belgium, 1.0	6,121		
			(14)Denmark, 1.0	7,052		

Source: OECD (2011: 28–29).

either energy efficiency in buildings and lighting, nor renewable energy, based on these as a percentage of total patents. In fact, Greece had a higher percentage of patents in energy efficiency in buildings and lighting as a percentage of total patent applications than the US.

Institutions influence *preferences* that order the processes by which policies are developed and implemented, and economic agents act. Nevertheless, there is clearly a difference in the intensity of climate innovation between more coordinated and more liberal states that may be set side-by-side with the difference in the intensity and application of policies and measures as demonstrated in Table 2.2. The LMEs have been far more interventionist than the CMEs across the board in addressing GHG emissions. While it is the case that Table 2.2 demonstrates that the latter have relatively leaned more heavily on market mechanisms and subsidies in an effort to meet their Kyoto commitments, what is perhaps most pertinent is that, across the board, the LME states have been more interventionist, especially in non-market policies and measures. The LMEs have around twice as many policies and measures as the CMEs but in respect of public investment and research and development support in particular, as well as for what might be regarded as market enhancing initiatives in respect of education and outreach and voluntary initiatives, they have three to eight times as many policies and measures. Therefore, although governments in LMEs also have more market based policies and measures, they appear to have a predilection for greater regulation through legislation and formal regulations to address what is regarded as market failure, by comparison to CME governments.

Block (2008b) and Block and Keller (2009, 2011) have investigated the paradox that the more firms become market focused, the more the state must act in respect of technological innovation not immediately dictated by the market in the short term. As Pryor (2002: 363) puts it, increasingly liberal market relations ultimately diminish 'entrepreneurship, technological cooperation between companies, and the degree to which managers take their responsibilities seriously'. Therefore, the more economic relations approach the LME ideal, the more we should expect a growing coordinating role of the state to underpin this. Rather than LMEs being treated as 'a residual category...mostly characterized in negative terms, that is, in terms of what they lack', they should be 'analyzed in terms of the alternative logic that animates them' (Thelen, 2001: 73 quoted in Howell, 2003: 107). In this case, states which are extremely liberal in some respects (for example, competition in markets to coordinate economic activity) require extensive coordination in

Table 2.2 Relative intensity of national policies and measures for renewable energy, 1999–2012

	Market Mechanisms		State Intervention and Coordination				Information and Voluntary Initiatives		Total[a]
	Financial Incentives/ Subsidies (incl. taxes)	Tradable Permits	Policy Processes	Public Investment	R&D	Regulatory Instruments	Education and Outreach	Voluntary Agreements	
LMEs> CME %	69	95	92	32	856	328	85	268	293	96

Source: Source: International Energy Agency's *Global Renewable Energy Policies and Measures Database*, http://www.iea.org/textbase/pm/index_effi.asp, accessed 20 February 2012. States categorized as per Hall and Soskice's (2001a) typology.

others to balance and support this (for example, technological innovation for which there is no immediate market imperative, such as climate change mitigation). In the case of climate innovation, as with technological innovation more generally, if it is not driven by immediate market imperatives, the state must play a leading role in moving it forward.

The above discussion illustrates some of the key ways in which national institutional variations suggest different paths to climate innovation, and the introduction of goods and services embodying new technologies. These should be the focus for study rather than universal claims about the 'ideal' policy choices and approaches to take to mitigate climate change. This is because the institutions that underpin national economic systems serve an economic coordination purpose and are the product of social and cultural preferences as well as a response to the way in which markets function (Polanyi, 2001[1957]). Thus, while scientific and technological knowledge, entrepreneurship, and innovation are now broadly recognized generally as drivers of economic growth (Atkinson and Hackler, 2010), innovation emerges from complex national systems of knowledge, economic capacities, political choices, and education systems (Metcalfe, 2007; Metcalfe and Ramlogan, 2006; Gault and Huttner, 2008). Thus, knowledge, entrepreneurship and innovation are deeply embedded in a wide range of institutions that are discussed in the next section. These institutions not only coordinate economic activity, they coordinate it in a manner regarded as 'appropriate' in distinct national contexts (see also March and Olsen, 1989, 1998). Putting it simply, formal institutions are codified expressions of the national culture and the social preferences to which these give rise. In our theoretical framework, we propose four institutional dimensions along which a national innovation system may be measured for their effectiveness in supporting climate innovation: patient finance, management autonomy, acceptance of uncertainty, and 'climate collectivism'.

Patient finance

Because demand pull is much less effective, science push becomes more important. This means that financing for R&D must be more patient, prepared to invest in longer-term, higher risk innovation strategies with uncertain technical or market outcomes. As noted in Mikler and Harrison (2012), perhaps only 50 percent of necessary GHG emissions reductions will come from current technology or its incremental evolution. The balance will require radical innovations, some of which cannot yet be predicted, that will be driven by scientific insights and long-term development underwritten by financing arrangements that do not demand short-term investment returns. Yet, over time, a liberal ideology leads firms to be more reluctant to undertake basic research and long-term development. For example, much of the R&D now performed by US commercial ventures is in information technology (IT). In IT (as in many other sectors) the majority of the investment has been in incremental improvements with expectation of a short-term commercial payoff (The Economist, 2007).[4] At Microsoft basic researchers are expected to integrate their work into products quickly. As the Chief Development Officer of Cisco once commented (quoted in The Economist, 2007): 'we might decry this on a public-policy basis, but at least as far as public markets are concerned it is a Darwinian world. You live or die by that'. As such a perspective gains traction, the state must step in to fill the gap between the innovations necessary and what the market currently demands. For example, in the US long-term research is increasingly funded by the US government. As outlined in Mikler and Harrison (2012), following Block (2008b) and Block and Keller (2009), as US business has embraced an increasingly liberal ideology that it shares with similar states where a greater orientation to market signals is embraced, fewer private resources are invested in the basic research needed to drive radical innovations.

Research shows that institutions (often related to tax policy) that support effective venture capital are critical parts of an effective national innovation system in which market demand is clearly signaled or reasonably assured. Although it is true that states with more liberal economic orientations have historically had the best funded and most active venture capitalists (VCs),[5] recent research suggests that VCs in these states are much less effective at directly supporting innovation than is the conventional wisdom. Hirukawa and Ueda (2011) find that in reality the role of VCs is generally limited to short-term financing of the process of matching (primarily incremental) innovations to market demand: venture capital is 'for start-ups and early-stage financing of new businesses in high growth industries' (Tylecote and Visintin, 2008: 82). In the US the classic examples are VC investments in Internet companies like Apple, Google, and Facebook or 'apps' developers like Zynga. However, VCs (and private equity) have an investment horizon of a few years as their investors expect a tangible return within, usually, three to five years.[6] Even then the effects on innovation 'depend on informal but powerful cultural constraints' such as uncertainty avoidance and collectivism, concepts that are discussed in more detail later (Li and Zahra, 2012).

R&D is increasingly undertaken by firms but not always financed by them. With the notable exception of the US, R&D is more likely to be performed in the private sector in the CMEs (70–77 percent) than in the LMEs (42–62 percent) (Mikler and Harrison, 2012). However, a greater proportion of CME firms (68–77 percent) also provide the financing for their R&D than firms in the LMEs and ambiguous countries (41–65 percent). In other words, US firms perform a majority of R&D but do not pay for it.

In addition to the role played by demand in product markets, the role of financial markets varies between states. US capitalism has been labeled 'stockmarket capitalism' (Dore, 2000) because of US firms' reliance on equity finance. Indeed, stock market capitalization is of a magnitude two to three times greater in the US by comparison to countries with a less liberal institutional foundations to their economies, such as Germany and Japan (Hall and Soskice, 2001a: 18).[7] This means US firms' access to finance is contingent on shareholder approval, which in turn is contingent on assessments of publicly available financial data and payment of financial returns in the form of dividends. Because US capital markets are highly liquid, shareholdings are volatile and changes in market sentiment can lead to rapid changes in firms' ownership. Because shareholder value is their primary goal, firms must adopt a short-term, shareholder-focused strategy or risk being starved of the capital they need to invest and survive (Vitols, 2001). Ironically, technological innovation has increased capital market volatility – and, thus, possibly discouraged long-term investment in productive innovations – by enabling 'high-frequency trading that comprises some 60 to 75 percent of all US market trades and multiple derivatives markets, some of which caused the US financial debacle in 2008.[8]

In Germany and Japan debt finance has been more the norm. Banks are often represented on the boards of major German companies, and are regarded as strategic industry partners rather than simply financiers.[9] In Japan, the major financial groups are often attached to, or closely affiliated with, large corporations. It has traditionally been common for large firms to rely on one bank for all their capital requirements (for example, see Hollingsworth, 1997). More than half the equity of Japanese firms is held by 'stable shareholders': banks, insurance companies and related companies with which the firm trades or has joint ventures (Dore, 2000).[10] Therefore, rather than being monitored by shareholders on the basis of their short-term financial performance, firms in countries such as Germany and Japan which rely more on debt finance also rely more on trust and support from their stable shareholders and financial partners based on their reputation, and for strong financial performance in the longer term (for example, see Hall and Soskice, 2001a). As such, they are said to embody 'stakeholder capitalism' (Dore, 2000). This means they are more 'immune' to short-run capital market fluctuations than LME-based US firms which must be more focused on short-term profit maximization. As such they cannot afford the luxury of incurring the disapproval of equity investors for long (Hollingsworth, 1997: 293).

In summary, climate innovation seems to demand patient finance, long-term investment in uncertain scientific research and early product development, unaided by clear market demand signals. Development and diffusion of climate innovations, especially more radical innovations, is likely to take several years and market demand is minimal and unpredictable. Therefore, equity markets that demand short-term returns are an inadequate source of the necessary financing. More stable and patient financing through debt or from single powerful shareholders, or through monopoly market positions, permit the longer-term strategy that climate innovation demands.

Management autonomy

Firms in which management has greater autonomy from shareholders should be better able, and therefore more likely, to invest in uncertain and costly climate innovation. The more autonomy that management has, the better it should be able to make a long-term, uncertain investment in climate innovation. The short-termism that is particularly common in shareholder market-driven LME-based corporations mitigates against long-term investment in GHG mitigation technologies. Firms will invest in radical climate innovation only if they have a long-term strategic view of business opportunities, and as the tenure of US chief executive officers has shortened, and they have become even more focused on short-term stock performance, so have firms' investment horizons (Antia et al., 2010). Changes in corporate governance rules to reward longer-term investment require coordinated policies on pay and incentive schemes, taxation, and employment tenures, and remains controversial and politically difficult (Jackson and Petraki, 2011).

Stakeholder theory argues that everyone with whom a firm has contact from vendors, employees and their representatives, local communities in which they do business, to shareholders have a legitimate interest in how the firm operates and what investments it makes (Freeman, 2010). Such theoretical perspectives suggest that social responsibility is assigned to corporations (that is, corporate social responsibility or 'CSR') as demands for social and environmental actions and reporting restrain corporate management from being purely profit-focused by being, or made to be, accountable to a broader variety of stakeholders. Tylecote and Visintin (2008) show that a broader range of institutions that regulate corporate governance influence firm risk-taking just as those that regulate finance described above do. For example, the finance and governance of LME firms are 'outsider-dominated' while CME firms are more 'insider-dominated' (Tylecote et al., 2002). The insider-outsider dichotomy is replicated in state policymaking: in CMEs, business, labor and environmental groups are formally included in policymaking, whereas in LMEs they are usually excluded, especially in environmental matters (Dryzek et al., 2003). As such, managers in outsider-dominated LME firms are less constrained and may assume greater business risks such as investing in high growth industries (Morck and Yeung, 2009), but this does not mean that environmental concerns are as easily internalized as is necessary for climate innovation.

'Corporate governance' comprises the laws, regulations, and self-regulation through codes of conduct, standards, and best practices that influence every aspect of firm ownership and operations, including directors' responsibilities, dissemination of financial information, relations with employees, taxation, and executive's incentive pay. In short, firms are systems that are open to all the national institutions that determine the distribution of power among stakeholders. These institutions are the result of national history, 'shaped by a form of corporate governance plate tectonics, in which the demands of current circumstances grind against the influence of initial conditions' (Gilson, 1996, quoted in Gilson, 2006). For example, the institutions of corporate governance in the US have changed as a result of economic forces (increased takeover activity in the 1980s), activism by institutional investors, and the later rise of equity-based pay; from managers persuading states to allow anti-takeover provisions like poison pills; and from federal imposition of greater regulation in response to the Enron debacle (Jackson, 2010). Managers of firms with a single stable, controlling shareholder – that is more often the case outside the US and UK and more common in CMEs – are more risk averse, because the controlling shareholder may be less able or willing to assume large financial risks and are able to prevent management from taking them (Gilson, 2006). Thus, they are commonly expected to prefer investing in incremental innovation and vertical integration.

This is a contested and complicated area, and one that needs greater study in general, as well as specifically in relation to climate innovation. For example, although it may be argued that firms that are more reliant on debt finance and often have a controlling shareholder – such as a financial institution as in many European and East Asian CMEs – may take a longer-term view than US firms, it may equally be argued that as US managers have become more insulated from stockholders than even managers in other LMEs, such as the UK, they may assume greater financial risks (Aguilera et al., 2006; Morck and Yeung, 2009). Equally, the structure of financing and corporate governance rules often pull in different directions. Firms that rely on capital markets and a broad shareholding may be less inclined to take the long view because of the need for short-term returns to investors but their greater insulation from shareholders' needs allows managers of US firms to take larger risks. However, this risk-accepting culture may not support investment in long-term technological innovation – like radical climate innovations – without a clear market opportunity that usually is absent. On the other hand, controlling shareholders in CME firms may be risk averse but more patient than equity markets.

The degree of competition in a firm's industrial sector is also a factor. Firms with a monopoly or near monopoly position in their sectors may choose long-range investments to maintain barriers to entry (as AT&T did for many years with Bell Labs) or they may abjure innovation especially if their dominant position is buttressed by political choices.[11]

Voluntary and formal CSR initiatives – that are a weak indicator of corporate concern for the natural environment – are recognized as an increasingly important contribution to corporate governance, especially in LMEs. However, managers generally see economic imperatives and CSR as separate objectives. For example, in a cross-national study of managers in nine nations, in only two countries (China and Denmark) did managers perceive compatibility between economic and CSR goals (Usunier et al., 2011). And the social role of CSR is seen differently across national borders. In LMEs, CSR is primarily perceived as a competitive tool but in CMEs it is seen as 'predominantly socially cohesive' (Kang and Moon, 2012: 101). Yet, some research argues that, especially in LMEs, CSR can be a voluntary effort by management to fill voids in formal institutions (Kang and Moon, 2012; Jackson and Apostolakou, 2010). But although the Jackson and Apostolakou (2010) study does not include the US, the greater insulation of US management from outside stakeholders beyond shareholders suggests an incentive for US managers to use CSR strategically without broadly adopting sustainable practices. Because institutional investors in the UK have greater influence over company policies than in the US, they are more able to influence how seriously CSR is treated: much more seriously in the UK than in the US (Aguilera et al., 2006).

Acceptance of uncertainty

Firms investing in climate innovation, especially more radical innovations, have to be able to finance long-term investments with very uncertain returns. Because it is likely to emerge more from scientific advances than market demand, investment in climate innovation is much more uncertain than investment in more conventional technological innovation. High-impact entrepreneurship – of which climate innovation could be an example – would appear to depend on social acceptance of entrepreneurial opportunities and acceptance of uncertainty and risk-taking in specific national contexts (Stenholm et al., 2013).

Scholars have found it extremely difficult to directly measure culture and innovation and to identify a correlation between the two concepts. In these studies, culture is generally accepted as 'the collective programming of the mind which distinguishes one human group from another...the interactive aggregate of common characteristics that influence a human group's response to its environment' and several aspects of culture have been suggested as potentially influencing innovation (Hofstede, 2003).[12] Of the cultural indicators tested by Scott (1993), acceptance of uncertainty was most strongly correlated with innovation. Other factors tested were much less strongly correlated or had 'no explanatory power'. Despite the weaknesses in indicators and absence of some data, Scott concluded that 'culture matters...highly innovative societies have people who are individualistic, low in power distance, and *accepting of uncertainty*' (emphasis added).

A recent analysis using different indicators, and datasets that were not available two decades earlier, reached a more nuanced conclusion. Taylor and Wilson (2012) confirm that a culture of individualism strongly correlates with cross-national indicators of innovation and find that several economic factors such as openness to trade, military spending, or (negatively) natural resource endowments have little influence on innovation. The authors also note a strongly positive correlation for intellectual autonomy – an aspect of individualism – but only under limited conditions. Scholars normally relate individualism to the number and social acceptance of (and rewards to) innovators, in other words to the supply of innovators that is presumed to directly relate to the supply of innovations. Yet, individualism may have a stronger effect on the *demand* for innovations either because consumers seek to customize technologies to their needs or because firms identify and satisfy unexpressed needs for individualized uses of technology. As market demand for climate innovations is likely to be low for the foreseeable future, more culturally individualistic nations can be expected to have no advantages over more collectivist nations.

Climate collectivism

Comparative capitalism scholars controversially assert that more liberal (and individualistic) forms of capitalism tolerate uncertainty better and are, therefore, more radically innovative than more collectivist and coordinated forms (for example, see Taylor, 2004). The literature on technological innovation is dominated by economic theory and, therefore not surprisingly, individualism. Yet, the rapid economic advance of more collectivist East Asian states, especially China, calls into question the widely-accepted hypothesis that individualism is a major cultural force driving innovation. While there is no evidence that collective 'familism' and 'localism' can drive technological innovation – and there is some evidence that it may harm science – Taylor and Wilson's (2012: 235) analysis suggests that 'a certain type of collectivism (that is, patriotism and nationalism) can foster innovation at the national level'. Institutionally collective cultures might 'produce a social environment in which both innovators and those bearing the costs of change are more willing to endure...difficulties for the benefit of their society'. Therefore, they conclude that 'it may be that nationalism is comparatively better at fostering technological change, while individualism is better suited to aid scientific inquiry'. While it is difficult to measure nationalistic collectivism, the limited available evidence seems to suggest that this tentative hypothesis is worthy of further investigation (as it is in Chapter 9 in the analysis of Chinese versus US corporate motivations for climate innovation).

Cross-national studies such as Hampden-Turner and Trompenaars (1993), have found that managers from firms based in states embodying non-liberal forms of capitalism are less driven by price signals in markets and act more on the basis of deep relations formed over time with clients, other firms, and the state. For example, they note that managers of US firms believe that 'if they are profitable, then everything else must be all right'. By comparison, 'for Germans, value must be deeply imbedded in products of solidity and worth [because] they do not like it when money and its enjoyment becomes separated from worthwhile artifacts' (Hampden-Turner and Trompenaars, 1993: 44 and 213; more recently in respect of Germany see also Streeck, 2010).

For nationalist collectivist culture to drive climate innovation it must be appropriately 'green'. That is, public opinion and public policy will indicate a high level of support for mitigation of climate change. The nation, including its firms, will expect that something should be done and are prepared to support effective action. As noted above, and in Mikler and Harrison (2012), continental European CMEs have generated more climate innovation even while the US outscores them on several common measures of innovation. Also, the CMEs have continued to lead in international negotiations even as several member nations fought to avoid sovereign default during 2011–2012, indicating a broad acceptance among their publics and political elites that climate change is a problem and that mitigative action should be taken. Finally, environmental interest groups and social movements are more likely to have a 'seat at the table' and a voice in the debate in CMEs than in LMEs. Such 'ecological democracy' appears more developed in European CMEs than in the US (Sousa and McGrory Klyza, 2007; Dryzek, 1997).

This may be a key reason why there are clear differences in actual changes in GHG emissions between LMEs and CMEs over the period 1990 to 2011. Excluding the UK, where emissions decreased by 28 percent,[13] the LMEs of the US, Ireland, Australia, Canada and New Zealand all increased their emissions, and by 17 percent on average. Excluding Iceland, which greatly increased its emissions by 26 percent, the CMEs of Austria, Belgium, Denmark, Finland, Germany, Japan, The Netherlands, Norway, Sweden and Switzerland decreased their emissions by 8 percent on average, while for the European Union (EU) as a region there was a 15 percent decrease. This dichotomy between LMEs and CMEs seems *prima facie* evidence of a greater willingness by CME consumers and their firms to embrace goods and services that embody lower emission technologies, as this has occurred despite fewer policies and measures being in place, less public expenditure on R&D, and similar market conditions (more positive before 2009, and less so afterwards). The LMEs, with the notable exception of the UK, have failed dismally in meeting their GHG reduction commitments, while the

CMEs have on average met theirs, and in some cases greatly exceeded them (UNFCCC, no date b).[14]

None of this *proves* that the US will be a laggard in climate innovation until public pressure rises significantly and we stand by our prior argument that LMEs are likely to be less effective generators of *radical* climate innovations than more cooperative and coordinated economies. Still, it is reasonable to hypothesize that countries that are more united in their belief that climate change is a real and present danger and accept greater coordination of economic activity toward a common goal of significant reductions of GHG emissions are more likely to generate climate innovations. The leadership of the European Union in international negotiations and the intransigence of the US, and the greater relative investment by European firms in climate innovations than the US both suggest that climate collectivism not only can drive a unified national search for climate innovations but may be a necessary if not sufficient condition for substantial mitigation of GHG emissions through technological innovation. Yet, without a fully comparative approach using data from several LMEs and CMEs this hypothesis cannot be properly tested.

Conclusion: Measuring institutional capacity for climate innovation

National institutions are enduring and key *determinants* of corporate organizational structures and choices by influencing the social technologies that they adopt. Regardless of whether national institutions functionally serve their purpose most efficiently or not, and regardless of whose interests they serve, once in place they endure and frame future processes of change (for example, see Mahoney and Thelen, 2010 and Olsen, 2009). This does not preclude change but channels it and limits it. A key insight of the comparative institutional scholarship is that even in the face of exogenous shocks, and endogenous pressures for change – and recognizing that capitalism is characterized by 'tumult' (Sewell, 2008) – the historical embedding of institutions produces 'a politics of institutional stability' (Hall and Thelen, 2009: 12) rather than flux. However, at critical junctures equilibrium may be punctuated by a radical shift in institutional arrangements, usually orchestrated by government actions in response to a perceived 'crisis'.[15] For example, in response to the financial crisis in 2008, the US Congress enacted a more than thousand page Act[16] that attempts to reform the financial system and nearly two years later is still being converted into specific regulations that affect nearly every part of it. In response to the 2011–2012 potential default of Greece and other European nations, the European Union organized a new financial stabilization fund and its political leaders demanded revisions to the Lisbon Treaty to limit national deficits and grant the EU sanction powers.

Climate innovation is not simply a question of assessing whether or not there is political will on the part of a government to address the problem of mitigating climate change, nor simply the stringency and enforcement of its regulations, nor the level of social concern, nor the economic incentives. It is not just a matter of getting the prices right in order that there are incentives for firms and consumers. Nor is it just a matter of optimality or efficiency in the way economic systems address the problem. It is our contention that it is a matter of getting the institutions right. At its most fundamental level climate innovations will emerge from national innovation systems constituted of those institutions that support appropriate innovation by profit-seeking firms. As we have shown, climate innovation demands a different national innovation system from the current, market-oriented innovation systems.

As shown in Table 2.3, we hypothesize that effective climate innovation requires high ratings on all four dimensions. Individual countries may be subjectively assessed on each dimension against the

Table 2.3 Hypothesized national measures of climate innovation

	'Good' Climate Innovation	Subjective Assessment of the US	Idealized CMEs
Patient Finance	High for private financing	Low for private financing; medium for government financing	High for private financing with controlling shareholder or debt financing
Management Autonomy	High	High	Low if controlling shareholder
Uncertainty Acceptance	High	Medium	Low
Climate Collectivism	High	Low	High

characteristics of the optimal climate effective national innovation system. For illustration purposes we show our subjective measures of the US compared to an 'idealized' CME country.

While grossly simplified, our subjective scoring in Table 2.3 reflects the discussions and data throughout this chapter. Taking the US as an example, firms' reliance on capital markets means that private financing is impatient and venture capital and private equity provide minimal support, at best, for long-term, science-based, technological innovations in the absence of perceptible market signals. This short-termism is supported by management's equity compensation and the shortening tenures of CEOs. Yet, management autonomy in US firms is high because of dispersed shareholdings and favorable corporate governance rules, except when occasional bursts of shareholder activism temporarily limit them (as after the failures of Enron and World Com). US culture supports and rewards risk taking but entrepreneurial activity may be less effective, because it is less patient for climate innovation than for market innovations. Hence, our subjective score of 'medium' for uncertainty acceptance. The US prides itself on an individualistic culture that is opposed to collective action, in which consumers and firms are expected to pursue their individual preferences, whereas action on climate change requires the capacity for collective action to mitigate its impacts.[17]

From our analysis, we identify four paradoxes that may impact clear institutional guidance for climate innovation. These are as follows:

- First, as discussed, LMEs require more intervention and a greater and growing role for the state to drive the climate innovations than is the case in CMEs. In the US, the lack of patient private finance means that the federal government has to substantially fund most scientific, and much other, research both directly and through grants and contracts with firms and universities.
- Secondly, CME firms with controlling institutional shareholders or a close debt relationship with a large financial institution have more patient finance than the US or most LMEs. Yet, those relationships tend to limit management autonomy. Exactly how those tensions play out will depend on their context. For example, if climate collectivism is strong, there may be a communion of minds between the providers of finance and management.
- Thirdly, cultures that are less accepting of uncertainty tend to be more collectivist, while cultures that are more individualistic better accept uncertainty. However, as we have argued, the uncertainty of climate innovation is distinct from that 'normally' encountered by

firms. Although they are usually expected to be more tolerant of uncertainty, low market demand for climate innovations *per se* will create greater uncertainty for firms in liberal capitalist economies. They are accustomed to innovating to directly satisfy market demand, whereas firms in CMEs should be less affected by a paucity of market demand, especially if a broad range of stakeholders participate in strategic decisions and provide an impetus for climate innovation.
- Finally, the nature of government intervention may be in opposition to the preferences of economic and social actors in distinct national contexts. For example, the more there is a preference for markets as coordinators of economic activity, the less there will be a desire to interfere in them. Societies, governments and firms in more liberal economies will prefer direct regulation, voluntary commitments, consumer education and subsidies for research that is not market driven, rather than interference in markets through taxes and trading schemes. The latter suit an institutional environment more comfortable with state coordination and a lesser role for markets as a 'place' where the business of business is more purely business (to paraphrase Friedman, 1970).

These paradoxes show that all four dimensions interact and that to generate more climate innovation requires modifying the whole national innovation system to accommodate its special needs. Improvement in no single dimension will suffice to increase national climate innovation.

Our analysis also suggests that the mechanisms of change in national innovation systems will differ between nations. The more informal and socially embedded nature of coordination in CMEs, means that less recourse to formal rules and regulation is preferred. Generally, the more that formal government intervention is required in more liberal institutional contexts, the more the need for direct and formal regulation rather than leaving the market and market actors to their own devices. For example, if consumer demand primarily drives corporate strategy in LMEs, while more reflecting corporate strategy in CMEs, then social attitudes are important for LME-based firms but only if reflected in material market outcomes, while CME-based firms will consider such attitudes and alter market outcomes in the process of altering production strategies and, thus, guiding the market. Therefore unless social concerns can be translated into consumer demand or government regulations in LMEs, they will have less of an impact in driving innovation and the introduction of new technologies in more liberal contexts as opposed to more collectivist ones.

Notes

1. Data accessed at http://www.recovery.gov/About/Pages/The_Act.aspx on 10 May 2012.
2. For definitions of 'incremental' and 'radical' innovations see Mikler and Harrison (2012).
3. We use LMEs and CMEs not because we uncritically accept the veracity of the VOC Approach of Hall and Soskice (2001a) who coined the categories, but as shorthand for states that may simply be thought of as more liberal and market coordinated, versus those that may be thought of as more collectivist or state-directed, and therefore less market coordinated. That is to say, corporations based in more liberal capitalist states prefer to coordinate their activities via market competition more 'arms-length' in their interactions with the state and more focused on consumer demand and providing shareholder value. By comparison, CME-based firms prefer more non-market cooperative relationships to coordinate their activities, including with the state and society.
4. US multinational corporations have either closed or moved many of their laboratories overseas. Between 1945 and 1979 Bell Labs produced ten radically new technologies that have changed the world. Its past successes are unlikely to be replicated today, because in 2007 Bell Labs had 1,000 employees by comparison to five decades earlier when it had 25,000. Today, IBM has research locations in eight countries and General Electric in five. Both companies have major R&D facilities in India and Brazil.
5. In 2009 information technology venture capital funding in the EU was less than venture capital funding in Silicon Valley in the first three months of that year; some of the largest European venture capital firms are in the UK (The Economist, 2010). Also see Table 3.11 in Tylecote and Visintin (2008: 83).
6. Private equity, much in the news because the presumed Republican contender for the US presidency (Mitt Romney) made his millions with Bain Capital one of the first Private Equity firms, used to be called 'leverage buyout shops' or 'asset strippers'. They collect pools of money from investors to buy companies, improve their efficiency, leverage their investment with debt secured by the purchased company's assets, extract huge fees and dividends, and sell them for a large profit either to another company or through a public offering. They have no interest holding the company for an extended period and little interest in investing in research and development. Their approach mirrors Bain Capital's roots: it was spun out from their consulting operation and focused on wringing inefficiencies out of firms' operations.
7. This is despite per capita GDP being comparable between LMEs and CMEs on average, data for which is provided on p. 19 – that is, the magnitude of stock market capitalization in the US versus CMEs such as Germany and Japan is not a function of the different magnitudes of their economies.

8   Estimates are summarized at http://www.zerohedge.com/article/what-percentage-us-equity-trades-are-high-frequency-trades, and Nina Mehta, 'High-Frequency Firms Triple Trades Amid Rout, Wedbush Says', Bloomberg 8/12/2011 at http://www.bloomberg.com/news/2011-08-11/high-frequency-firms-tripled-trading-as-s-p-500-plunged-13-wedbush-says.html. Both were accessed on 17 May 2012.
9   A point made repeatedly in the comparative capitalism literature more broadly. See for example Dore et al. (1999), Wilks (1990), and Pauly and Reich (1997).
10 See also Pauly and Reich (1997: 10) who point out that cross-shareholdings in Japanese firms have barely changed in past decade.
11 Examples include 'national champions' such as Russia's GazProm, France's Areva, Mexico's América Móvil, and China Petroleum. An open economy with domestic competition helps firms compete internationally.
12 The first edition of Hofstede's book was published in 1980 and became a seminal text in this literature. The relevant characteristics of culture were also described in his study. Criticism of Hofstede's work has revolved around both methodological issues and epistemological issues, and the use of a survey of 80,000 IBM employees across the world. See, for example, McSweeney (2002).
13 Of course, its GHG emission reduction commitments and efforts at meeting them are complicated by being a member state of the EU.
14 For example, Germany's GHG emissions fell by 27 percent.
15 The concept of punctuated equilibrium was introduced by Eldredge and Gould (1972) and adapted to social theory by Baumgartner and Jones (1993). It was further adapted as the concept of 'self-organized criticality' to complexity science by Bak (1994, 1996). All three approaches essentially posit that systems remain in equilibrium and only evolve in occasional spurts. In paleobiology the unexpected eruption of new life forms seen in the Burgess Shale was explained as the result of some radical change in external conditions. In complexity science an emergent system self-organizes into a critical state in which sudden and rapid changes can occur as a result of the interaction of internal agent relations (from the self-organization of the system) with external stimuli.
16 Dodd-Frank Wall Street Reform and Consumer Protection Act (Public Law 111–203).
17 Of course, this is simplified and idealized view as cultural norms and fashions influence the behavior of all economic agents.

Part I
The National Context of Climate Innovation

The current environmental predicament in which the world finds itself demonstrates all too clearly that the technology systems of *all* capitalist economies have not generated sufficient climate innovation. The world is facing the threat of climate change *because* national innovation systems have been oriented to growth in jobs and output rather than reduction of greenhouse gas (GHG) emissions or increases in energy efficiency. To prevent dangerous climate change, nations have to redirect their innovation systems to produce technologies to reduce future emissions and, if possible, to reverse the harm caused by the products of current systems of innovation. To understand how to increase the quantity and quality of climate innovation in these systems requires analysis of their national institutional contexts representing different varieties of capitalism. In this book we focus on liberal capitalism, and specifically on the United States (US) as the exemplar of that mode of political economy. In this part, comprising two chapters, we begin to add some flesh to the bones of our theoretical framework by investigating the general properties of its national innovation system.

The US is supposed to be the archetype of liberal capitalism, but in Chapter 3 Robert MacNeil shows that there is a dense network of research and development (R&D) programs and institutions that have grown into a developmental network state (DNS) apparatus that underpins the US's remarkable track record in technological innovation. With a detailed analysis of the vast array of legislation, programs and agencies involved, he demonstrates that this DNS apparatus has been turned towards technological innovations to reduce greenhouse gas (GHG) emissions, primarily from energy use. However, he shows that market and economic considerations still dominate these programs. While ostensibly existing to promote and enable climate innovation, the US's DNS is actually more an enabler of the normal market innovation process outlined in Chapter 1. There is no overt strategic rationale of meeting emission reduction targets, and economic concerns such as reducing companies' costs or risks retain primacy. While scientific panels are used to select targeted technologies, financial assistance is geared to economic concerns rather than reductions in GHG emissions. Thus, he demonstrates that there is a developed architecture for climate innovation but that this formal system lacks the institutional structure it would need to generate climate innovation. Ultimately, the institutional basis of the US's DNS apparatus is primarily designed to achieve economic and political, not environmental, outcomes.

In Chapter 4, Stratis Giannakouros and Dimitris Stevis consider whether initiatives at the sub-national level may hold more hope for the climate innovation necessary to reduce GHG emissions. Their central point is that some states within the federal US system have adopted policies that look very much like versions of ecological modernization. By tracing the development of Colorado's New Energy Economy (NEE) from 2004 to 2012 they show that sub-national polities can encourage technological innovation that mitigates GHG emissions while also supporting desirable economic outcomes. Even within the US liberal market economy (LME) that is ideologically opposed to an activist state to drive technological innovation, it is possible for individual states to move in the direction of ecological modernization. They also show that even within an LME such as the US, states are not totally constrained by institutional path determinacy and may choose to construct an institutional ecosystem that effectively supports climate innovation. Unfortunately, they also show that sub-federal attempts at climate innovation, while important and necessary, are individually insufficient to address the collective action problem that climate innovation presents in a liberal context. A long-term solution requires more profound institutional innovations and strong coordination between federal and state levels.

In Chapter 1, we defined climate innovation as technological innovation with the primary purpose of mitigating climate change. What is demonstrated in these chapters is that, while the US innovation system nominally targets climate mitigation, in reality it continues to support market innovation. In the mode of you-get-what-you-pay-for, national funding that is oriented to market innovation cannot produce climate innovation, and climate innovation at a sub-state level is 'trumped' by economic motivations at the federal level. Therefore, in the US innovative technologies to mitigate GHG emissions are not an end in themselves but a by-product of economically motivated market innovation. This supports our contention in Chapter 2 that liberal capitalism militates against climate innovation because it is the institutions of liberal capitalism, rather than the policies, that determine the outcome. Without fundamental institutional change, the policies, no matter how worthy, are likely to be ineffective in sufficiently mitigating climate change.

3
Climate Policy, Energy Technologies, and the American Developmental State

Robert MacNeil

Over the past three decades, the United States (US) economy has produced an enviable record of high-tech innovation in industries ranging from information technology, to biotech, medical engineering, semiconductors, software, telecommunications, and defense. While this record has long been acknowledged and celebrated, attempts to understand the roots of this success have often been confounded by misguided conceptions about the institutions responsible for facilitating it. One particularly influential narrative (commonly referred to as the Varieties of Capitalism framework) has generally sought to attribute the American style and capacity for high-tech innovation to its supposedly liberal form of market economy, in which relatively unfettered markets provide the conditions necessary to promote rapid and disruptive technological changes across the economy. This chapter begins from the assumption that this general fetishization of market forces impairs our ability to properly understand the American state's role in high-tech innovation generally, and climate innovation specifically.

In this chapter, I demonstrate the extent to which, over the past three decades, American markets for innovation have been underpinned and supported by a dense network of federal R&D programs and institutions, the aggregate of which constitute a powerful and proactive 'developmental network state' (DNS) apparatus within the federal government (Block, 2008b). My primary claim (which I demonstrate empirically) is that Washington's efforts to promote and facilitate domestic climate and energy innovation are making extensive use of the very same DNS institutions and policies that have proved crucial in the success of other US high-tech markets since the late 1970s – thus belying the notion that American innovation is owed primarily to a dynamic and relatively uncoordinated private sector.

At a theoretical level, I argue that the presence of a DNS within the world's archetypal neoliberal state should not necessarily be understood as a paradox – nor is it necessarily at odds with the common portrayal of the US as a so-called 'liberal market economy' (LME). Rather, building on the tenets of Regulation Theory, I argue that the presence of a stealthy developmental state apparatus in Washington has emanated from the very tensions that capitalist states face when attempting to promote stable accumulation under conditions of neoliberalism. Put simply, the more that 'the sole business of business is business' under neoliberalism – for example, shareholder value, profits, and so on – the greater the need for an active regulatory state (consisting of institutions like a DNS) to stabilize and maintain the accumulation process. In this context, the DNS not only provides a massive hidden subsidy to the private sector, but also helps policymakers to establish stable 'modes of regulation' around national innovation.

On first inspection, the willingness of the US federal government to become a proactive facilitator of climate innovation may seem auspicious. Indeed, as Mikler and Harrison note in the opening chapter of this volume, in light of several uniquely prohibitive characteristics of climate innovation, the 'normal' market innovation process will not suffice in this case. Rather, the state will be required to exercise greater interventionist power all throughout the process, from the initial funding of basic research, through development, demonstration, and commercialization. However, I will argue in this chapter that despite their immense success in a range of other high-tech sectors over the past thirty years, the DNS institutions being deployed at present are largely ill equipped to promote successful climate innovation. Put simply, the American developmental state was never designed to launch markets based on political prerogatives like, for example, emissions reductions or environmental protections. Nor was it ever designed to challenge or undermine the hegemony of existing market participants (like, for example, conventional energy interests) or destroy existing technologies. It has historically functioned *only* to develop technologies aimed at meeting market needs not currently being filled by some other cheaper technology. A DNS strategy, in this context, must be backed by powerful regulations on fossil energy use – something which has proved yet impossible given neoliberalism's anti-regulationist bent. Green technologies emanating from the DNS are thus forced to compete in a stilted energy market that is hostile to their uptake. As this chapter will note, this simple fact has greatly limited the potential utility of Washington's climate innovation strategy.

Why does the US have a DNS? National innovation strategies as 'modes of regulation'

Before talking specifically about the nature of the DNS policies that have emerged around US climate innovation, it is worth briefly describing the tensions that have caused Washington to develop and maintain such an institution.

While the US has long been characterized as the world's archetypal LME (Hall and Soskice, 2001a), the characterization of American innovation as being a comparatively *laissez faire* market process fails to adequately capture the political-economic considerations brought to bear on national innovation policies. From a regulationist perspective, national innovation strategies can more appropriately be understood as a 'mode of regulation' deployed by states and policymakers in an effort to overcome market inconsistencies and crisis tendencies, and provide a tenable framework within which innovation cycles and economic accumulation can reproduce themselves. In this theoretical context, contrary to the supposed rationale of the LME model, markets are rather something to be *overcome* and offset through structured, political accommodation, and *not* left to self-direct in the hope that they will generate stable and consistent accumulation.

This theoretical framework is considerably more helpful for understanding the basic behavior of US innovation policy over the past half-century, and in particular the emergence of a DNS within the federal state. If one considers a) the state's dependency on stable growth and accumulation for its own material strength and social legitimacy and b) the role of continuous innovation in driving the accumulation process, the state's concern with actively encouraging innovation becomes much clearer. Historically, economies that have excelled at producing continuous technological advances have been able to reap immense benefits in the form of economic growth, lower unemployment rates, trade surpluses, higher taxation revenues for the state, and so on (Negoita, 2010). Britain provides an instructive example of this, as the massive array of technological innovations emanating from the country's rapid industrial growth throughout the 19th century ensured an uninterrupted domination of global markets for close to a century (Hobsbawm, 1999). In more recent times, the revolutionary transformation of Japanese auto manufacturers from mass production to flexible organization allowed them to cheaply produce smaller volumes of a wider variety of products, thus creating the conditions for the country to become a global leader in sectors like automotive production, and put an end to the US's post-war dominance of the industry (Brenner, 2002). The obvious historical lesson is that national economic growth thrives under conditions of continuous technological innovation, and states (which depend on economic growth for their own power and legitimacy) will tend to have a strong political interest in facilitating it.

It is this basic political and economic compulsion that led to the emergence of a DNS in Washington in the late 20th century. While Washington's early post-war innovation strategy had been predicated on the so-called 'pipeline model' of technological development – in which state investments in basic research technological innovation (primarily channelled through the defence and aerospace industries) were understood as making a natural transition to private consumer markets – by the mid-1960s the efficacy of this strategy had begun to wane. By this time, the immense economic lead and competitive edge the US had enjoyed since the close of the Second World War had begun to decline, leaving policymakers searching for more proactive policies for promoting innovation. The policy response, which began in the mid-1970s and reached a fever pitch throughout the Reagan years, would aim to hasten the transition of new technologies from government and university labs to the marketplace. In so doing, legislators would undertake the development of a series of new laws aimed at (1) coordinating the country's R&D efforts in a number of key sectors; (2) increasing the aggregate level of state funding for basic and applied research; (3) developing technology transfer mechanisms to ensure that publicly funded research found private investors to take it to market; and (4) creating specialized regulatory environments designed to foster these new technologies and markets.[1] The overall result of these legislative initiatives was the creation of what has been referred to as a developmental network state[2] in Washington. As Block (2008b: 187) notes:

> Taken together, these shifts have radically changed the national innovation system in the United States. A generation ago, a large portion of innovations could be traced to the autonomous and self-financed work that went on in the laboratories of Fortune 500 companies. Now, however, most innovation occurs among networks of collaborators that cross the public-private divide (2008: 11).

These developments correspond with the arguments of authors like Cerny (1997) and Palan et al. (2000) who argue that under conditions of intensified international competition, modern states – far from retreating, as neoliberal dogma suggests – become an increasingly important vehicle for achieving transformational economic goals and maintaining competitive industries. For Cerny, developments like this signal the birth of the so-called 'competition state', with policies responding primarily to economic pressures from rival competition states, all of whom are seeking to promote the viability of domestic industry and economic growth in a globalized economy.

A large body of empirical research has emerged over the past few years aimed at underscoring the crucial role of this developmental apparatus in fostering the growth of a range of high-tech industries since the 1970s (Block, 2008b; Block and Keller, 2009; Fuchs, 2010; Hurt, 2010; Negoita, 2010; Schrank and Whitford, 2009). Adding to this growing body of work, the following sections aim to demonstrate how the patterned deployment of these developmental state policies can be seen once again in the inchoate energy/climate context.

Climate/energy technologies and the Developmental Network State (DNS)

The emergence of DNS policies around climate and energy innovation date back as far as the late-1970s when the Carter administration, driven primarily by concerns about energy security, began making initial investments in a range of alternative and renewable technologies. This included the passage of the National Climate Program Act of 1978, the allocation of over $7 billion to R&D funding for alternative technologies, as well as the creation of a series of supporting agencies and programs, including the Department of Energy (DOE); the DOE National Laboratory System; the Office of Energy Efficiency and Renewable Energy (EERE); the Renewable Energy Resources Act; the US Synthetic Fuels Corporation Act; the Biomass Energy and Alcohol Fuels Act; the Energy Tax Act of 1978; the Energy Security Act of 1980; the Solar Energy and Energy Conservation Act; the Geothermal Energy Act; and the Ocean Thermal Energy Act (Simpson, 2003; MacNeil and Paterson, 2012).

While much of the funding for these programs remained in place throughout the 1980s, the early 1990s would see new life brought to these policies as the Energy Policy Act of 1992 included a large host of provisions aimed at spurring investment and employment in a range of alternative energy sectors (see Table 3.1 for select examples).[3]

Table 3.1 Energy Policy Act of 1992 alternative energy programs

Federal Initiative	Description
Innovative Renewable Energy Technology Transfer Program	A program to build foreign markets and buyers for domestic renewable energy technologies
Renewable Energy Production Incentive Program	A program to incentivize the adoption of renewable energy technologies by energy utilities
Renewable Energy Export Technology Program	A program to train workers in the system design, operation, and maintenance of renewable energy and energy efficiency equipment manufactured in the United States
Tax and Rate Treatment of Renewable Energy Initiative	A program to incentivize the timely development and deployment of domestic renewable energy technologies
Renewable Energy Advancement Awards Program	A system of incentives for groups promoting the advance of the practical application of biomass, geothermal, hydroelectric, photovoltaic, solar thermal, ocean thermal, and wind technologies to consumer, utility, or industrial uses
Alternative Motor Fuels Amendments	A program of incentives for groups developing alternative automobile fuels
United States-Asia Environmental Partnership	A program to build markets in developing Asian countries for environmental and renewable energy technologies developed in the United States
Coal Research, Development, Demonstration, and Commercial Applications Program	A program for research, development, demonstration, and commercial application on cleaner-burning coal-based technologies
Clean Coal Waste to Energy Program	A program of research, development, demonstration, and commercial application with respect to the use of solid waste combined with coal as a fuel source for clean coal combustion technologies
Clean Coal Technology Solicitation Program	A program to ensure the timely development of cost-effective technologies capable of reducing emissions from the combustion of coal

The Clinton years would see these efforts continue as the administration's Climate Action Plan established over forty federal programs for alternative and renewable energy R&D (Simpson, 2003; US Executive Office of the President, 1993). In addition to $22.2 billion in direct R&D funding through programs like the administration's flagship Climate Change Technology Initiative, these programs would include a series of targeted tax expenditures and loan guarantees totalling over $60 billion between 1994 and 2000 (US Office of Technology Assessment, 1995; US Department of Energy, 1995).[4]

Investments continued under the second Bush administration, which, in an effort to validate its position that action on climate change is better achieved through investments in technology rather than environmental regulation, assembled a vast climate research and innovation apparatus within the federal government comprised of the National Climate Change Technology Initiative, US Climate Change Research Initiative, Carbon Sequestration Leadership Forum, Climate VISION, and the Asia-Pacific Partnership on Clean Development and Climate. In addition, the administration oversaw the passage of the Energy Policy Act of 2005, the Energy Independence and Security Act (EISA) of 2007, and the Farm Bill of 2008 which, combined, allocated more than $12 billion in federal funding for biofuels, coal carbon capture and sequestration technology, wind, solar, geothermal, and tidal technologies, plug-in hybrid-electric vehicles, advanced battery development, and smart grid technologies (Stubbs, 2010).

Between 2009 and 2011, the Obama administration ramped-up these efforts with increased vigor. Beginning with the American Reinvestment and Recovery Act, the administration allocated $81.39 billion[5] to climate/energy technologies, followed by $30.73 billion[6] in its FY 2010 and 2011 budgets. In addition to establishing a series of new programs, tax expenditures, and energy innovation agencies, the administration further developed the second Bush administration's National Climate Change Technology Initiative, whose architecture mirrors the networking strategies that proved so successful in similar developmental state projects over the past three decades.

More instructive for understanding the nature of Washington's desire to promote climate and energy innovation, however, is the highly sophisticated networking apparatus that has been deployed to develop these new technologies. This decentralized system of funding and networking plays, in effect, four key qualitative roles in the innovation process, as it attempts to guide new technologies from the concept phase through to commercialization. These are described below as a) targeted resourcing, b) networking, c) technology transfer, and d) facilitation.[7]

Targeted resourcing

This initial step in the DNS process involves government officials consulting with experts in a given field (for example, defence, biotech, IT, aerospace, energy and so on) to identify crucial scientific and technological challenges, the solution to which would help advance the frontier of the industry, address certain practical state needs, and create opportunities for economic growth and accumulation (Block, 2008b). Developmental agencies relevant to each field then issue open solicitations to research groups describing the nature of the scientific problem they seek to surmount (or technological idea or concept they wish to develop), and begin providing funding and other resources to numerous competing groups with promising ideas. These solicitations can range from highly abstract basic research activities (for example, trying to understand the peculiarities of quantum mechanics or the nature of cells), to highly practical applied research activities (like developing new cancer screening technologies). Derived from the Defense Advanced Research Projects Agency (DARPA) model of innovation, this style of targeted resourcing can be distinguished from traditional bottom-up approaches to scientific funding, in which agencies distribute resources for novel inventions in the absence of explicit mandates. By contrast, creating mandates from the top-down allows policymakers to focus the efforts of the country's scientists and engineers on specific objectives, and thereby mobilize science in the direct service of the state's economic and security goals.

In the climate and energy realms, the targeting agenda is set by the labyrinth of agencies and working groups consolidated under the National Climate Change Technology Initiative (NCCTI). As Table 3.2 depicts, the initial policy agenda begins in the Office of the President (where very broad economic and security objectives are set), and moves down to the individual agencies responsible for executing the innovation agenda.

Responsible for setting the tangible scientific and innovation agenda, the Climate Change Technology Program (CCTP) is, in effect, where the rubber meets the road in the DNS process, as it creates a specific inventory of technology programs and works with the individual agencies in their respective efforts to implement their R&D activities.

Table 3.2 NCCTI targeting apparatus

Office of the President
↓
Climate Change Policy and Program Review
Operated by: National Security Council; Domestic Policy Council; National Economic Council
↓
Committee on Climate Change Science and Technology Integration
Operated by the heads of: Dept of Commerce; Dept of Energy; Office of Science and Technology Policy; Dept of State; Dept of Agriculture; National Economic Council; NASA; Dept of Transportation; Dept of Defense; Environmental Protection Agency; Dept of Interior; Council on Environmental Quality; Office of Management and Budget; Health and Human Services; National Science Foundation
↓
Interagency Working Group on Climate Change Science and Technology
Operated by the heads of: Dept of Energy; Dept of Commerce; Office of Science and Technology Policy

↓	↓
Climate Change Science Program Operated by: DOC, DOD, DOE, DOI, DOS, DOT, EPA, HHS, NASA, NSF, USAID, USDA	**Climate Change Technology Program** Operated by: DOE, DOC, DOD, DOI, DOS, DOT, EPA, HHS, NASA, NSF, USAID, USDA

Source: Climate Change Technology Program, 2006.

As Table 3.3 depicts, the CCTP assigns each agency a series of focus areas based on its expertise as it relates to energy, and requires them to develop their own R&D programs to develop capacities.

Targeting the private sector

Targeted resources are further provided to a range of private sector firms engaged in R&D efforts that dovetail with the objectives of the NCCTI. These activities operate at two primary levels. First, less-established companies (for example, small businesses, start-ups, and spin-offs) receive support from two federal programs called the Small Business Innovation Research program (SBIR) and Small Business Technology Transfer program (STTR). These programs mandate that all federal agencies with R&D budgets allocate a given percentage of that budget (2.5 percent under SBIR and 0.3 percent under STTR) to private sector innovation. The climate and energy realms feature the DOE as the key agency in this regard, establishing a new roster each year of individual program solicitations that invite small businesses to apply

Table 3.3 Selected developmental activities within individual agencies

Federal Agency	Examples of Energy R&D Activities	Developmental Program
Department of Agriculture	Soil sequestration, biomass energy, biofuels, cropping systems	Cooperative Research Grant Program; Sustainable Agriculture Research and Education (SARE); Bio-energy Research Service, Renewable Energy Program; Biomass R&D Program
Department of Energy	Energy efficiency, renewable energy, nuclear fission and fusion, carbon sequestration, basic energy sciences, hydrogen, electric grid infrastructure	Advanced Research Projects Agency – Energy; Energy Efficiency and Renewable Energy Program
Environmental Protection Agency	CO_2 mitigation, GHG emissions inventory systems, renewable energy	Environmental Technology Opportunities Program (ETO); Office of Research and Development (ORD)
Department of Defense	Aircraft, engines, fuels, trucks, power, fuel cells, lasers, energy management, basic research	Defense Advanced Research Projects Agency (DARPA)
USAID	Land use, sequestration, cropping systems	Energy Technology Development Program; Carbon Capture and Sequestration Program
Department of Commerce	Instrumentation, standards, ocean sequestration	Asia Pacific Partnership on Clean Development and Climate
National Science Foundation	Geosciences, nanoscale science and engineering, computational sciences	Energy For Sustainability Program
Department of the Interior	Sequestration, geothermal	Global Change Research and Development Program (GCRD); National Carbon Sequestration Assessment
Department of Transportation	Aviation, urban mass transit infrastructure, transportation systems, transportation efficiency	Research and Innovative Technology Program

Source: Committee on Climate Change Science and Technology Integration (2009).

for SBIR or STTR grants based on the priorities set under the NCCTI umbrella. Recipient firms are led through three phases of development, at the end of which they are expected to produce a functioning prototype that can be brought to market with the assistance of an array of DOE commercialization programs.

Over the past decade, however, there has been a growing recognition that programs like SBIR and STTR often fail to support technologies adequately enough for them to achieve commercialization. As a result, many innovations tend to collapse prior to reaching the market. Though a series of programs has been established at all levels of government seeking to remedy this problem, among the most radical and consequential has been the second major element of private sector targeting, the creation and proliferation of the modern 'public venture capital' (VC) firm at the federal level. As the name suggests, these entities are public VC organizations housed within individual government departments that use public resources to fund private firms engaged in certain types of R&D. The sudden expansion of the public VC model began innocently enough in 1999 when the CIA established a small investment company called In-Q-Tel with the goal of enhancing innovation and procurement efforts as defence budgets declined under the Clinton administration (Weiss, 2008). Within five years of In-Q-Tel's establishment, its positive reception had led to the creation of similar entities across a range of federal departments. As Keller (2010: 151) notes, today 'virtually every federal agency with a technology-focused mission has explored public venture capital as a means to stimulate technological innovation and/or commercialization of federal research'.

For its part, the DOE has been a leading agency in this regard. As of 2011, the department had launched four programs, including a venture capital partnership with the Battelle Memorial Institute; the Entrepreneur in Residence (EIR) program (through which VC funded entrepreneurs work with the DOE's national labs to identify and fund market-ready technologies); the Technology Commercialization Fund (through which monies are provided for prototype development, demonstration projects, market research, and other deployment activities); and a series of favorable loan guarantee programs through the DOE's Loan Programs Office (LPO) – the largest of which being the Advanced Technology Vehicle Manufacturing loan program and the '1703' and '1705' programs, which together have poured tens of billions of dollars into alternative technologies and the development of clean energy vehicle designs with major auto manufacturers.

The impact of these programs has been quite profound over the past several years, serving in many ways to reshape the broad nature of technological development in the energy sector. Acknowledging the extent to which federal funds can help attract a new private investment (and by the same token that falling outside the DOE's objectives can sink one's entrepreneurial hopes), many young companies engaged in developing and deploying various technologies have begun to tailor their business plans directly to the DOE and NCCTI's technology objectives in hopes of obtaining state assistance. For their part, private investors have awakened to this dynamic and are increasingly pulling back and waiting to see which companies will receive state assistance before making an investment. The result is that while the DOE had given out more than $18 billion in grants and loans to such firms by the end of 2009, private venture capital firms had invested a mere $2.68 billion (King Jr, 2009). Further complicating the equation is an obvious preference on the part of the DOE to invest almost exclusively in companies that have already procured substantial private funding, in the hopes that the role of federal funds can be limited to the creation of a 'critical mass' which can then attract increased private funding and thereby push the technology to commercialization. This has created a rather complicated 'chicken-or-egg' situation for many green energy firms, as private venture capitalists are reluctant to invest in companies that have yet to obtain state funding, while the DOE is wary of funding ventures that have yet to win over the private sector.

Networking

The state's second major role in the innovation process encompasses a wide range of efforts aimed at connecting the different laboratories, academic institutions, businesses, and research groups involved in the innovation process. This process establishes critical networks across the research community and provides scientists and engineers the opportunity to collaborate and exchange ideas while gaining crucial insights from others about critical problems and bottlenecks. Brokering further creates opportunities to put already existing technologies together in new ways, or allow one lab to combine a new technique from another lab with its own incremental change to render something completely new and different (Block, 2008b).

There are a number of informal hubs and liaisons capable of facilitating these critical brokering and networking activities, including officials and program officers engaged in targeted resourcing at individual federal agencies; individual university and national lab scientists at the centre of the university-industry collaborative research complex; technology transfer officials in the national lab system; a range of academic groups and agencies that provide a forum for disseminating ideas and discoveries both formally and informally; and an array of formal industry consortia through which private firms (backed by federal legislation) collaborate and share R&D secrets on pre-competitive research.[8]

In the climate and energy context, a range of initiatives have been established to facilitate these objectives, including the creation of Energy Innovation Hubs, Energy Frontier Research Centers, NSF Engineering Research Centers, and Industry/University Research Centers. For the most part, however, this process has been centralized under the CCTP's Integrated Planning and Networking program – an entity designed to help overcome incomplete knowledge and develop solutions to technical bottlenecks in the innovation process. To this end, the CCTP has established six multi-agency working groups aligned with the program's strategic goals. As Table 3.4 depicts, each working group maintains a series of sub-groups in which researchers from the participating agencies and affiliated labs are expected to share and exchange information relevant to the R&D process; coordinate inputs from all relevant agencies and systematically explore various technology program issues, gaps, challenges, impediments to progress, and opportunities; explore a range of potential research avenues to address the identified issues; and design a strategic research program to pursue the most promising avenues, including clear articulation of research goals and ideas for new solicitations of research proposals to address the identified areas (CCTP, 2006). These groups are further required to take part in the occasional, mandatory holding of technical workshops – events aimed at bringing together the applied and basic research groups from all agencies within the CCTP structure in a formal setting to exchange ideas, discuss bottlenecks and barriers impeding development, and discuss new research strategies. Through these forums, hundreds of individuals from research groups in the national lab system, academia, industry, and federal agencies are called upon to present and discuss new ideas and the current state of their research.

Technology transfer

Moving out of the lab and into the commercialization process, the state's third crucial role involves helping scientists and engineers make

Table 3.4   CCTP research and networking groups

Working Group	Sub-Groups
Energy Use Working Group – Led by DOE	• Electrical Grid and Infrastructure group • Hydrogen End-Use group • Buildings group • Transportation group • Industry group
Energy Supply Working Group – Led by DOE	• Hydrogen Production group • Renewable and Low-Carbon Fuels group • Renewable Power group • Nuclear Fission Power group • Fusion Energy group • Low Emissions Fossil Based Power group
CO_2 Sequestration Group – Led by USDA	• Carbon Capture group • Geological Storage group • Terrestrial Sequestration group • Ocean Storage group • Products and Materials group
Other (Non-CO_2) Gases Group – Led by EPA	• Energy and Waste group • Agricultural Methane and Other Gases group • High Global Warming-Potential Gases group • Nitrous Oxide group • Ozone Precursors and Black Carbon group
Measuring and Monitoring Group – Led by NASA Basic Research Group – Led by DOE	• Application Areas group • Integrated Systems group • Fundamental Research group • Strategic Research group • Exploratory Research group • Integrative R&D Planning group

Source: Climate Change Technology Program (2006).

the necessary business connections required to bring an innovation to market (or, alternatively, help an established business take ownership of novel technologies and processes developed in the national lab system). As noted above, assisting in the commercialization process became a central preoccupation of the state beginning in the late 1970s, as the pre-existing 'pipeline model' of technology transfer ceased to hold-up against the proactive industrial policies of Western Europe and East Asia. Under the Reagan administration, a tidal wave of new legislation was passed aimed at pushing public innovations to market, thus rendering the commercialization of new technologies the primary *raison d'être* of contemporary scientific inquiry (Block, 2008b), and making federal lab administrators de facto 'public sector venture capitalists' charged with moving novel technologies to the private sector as quickly and effectively as possible.[9]

In the climate and energy context, the most crucial point of technology transfer is the DOE national lab system – an organization of thirteen federally owned research facilities staffed by more than 30,000 scientists and engineers. The labs receive their R&D funding from Congress (and to a lesser extent private clients and research partners), which is then budgeted and prioritized by the DOE's Office of Energy Efficiency and Renewable Energy (EERE) and/or the private management company charged with operating the facility. Since technology transfer mandates were established, each DOE lab has been required to hire a full-time Technology Transfer and Intellectual Property Management Department staffed by 'professionals experienced in managing, marketing, and licensing intellectual property, as well as in patent prosecution' who orchestrate and oversee the transfer process, serving as a liaison between researchers and private firms and ensuring that inventions are subject to proper intellectual property protection (Lawrence Berkeley Laboratory, 2010).

Once the Technology Transfer Department identifies a novel invention developed in their facility that demonstrates commercial potential, the lab patents the technology and begins seeking licensees for it. If a suitable applicant is found, the department and applicant negotiate the terms of a licensing agreement which establishes the invention's issuing fee, the running royalty fee paid to the lab, and other financial terms relevant to the technology and market. A range of other transfer mechanisms (including the Work-for-Others program, Collaborative Research and Development Agreements, the Innovation Ecosystem Program, the Personnel Exchange Program, Technology Maturation agreements, and an array of User Facility Arrangements) also serve as key elements of the commercialization process.

Since taking on technology transfer as one their primary mandates, the DOE labs have been quite successful at bringing novel energy innovations to the market. In 2008 alone, the twelve DOE labs, along with the five other national labs, engaged in more than 12,000 technology transfer transactions. These included more than 2,500 Work for Others agreements, 2,800 user facility agreements, 700 Collaborative Research and Development Agreements, and more than 6,000 licenses for patented technologies. In addition, the labs reported more than 1,400 new inventions and filed close to 1,000 new patent applications (Perry, 2010).

Facilitation

The state's final role consists of attempts to create policies capable of building durable markets for the new technologies in question. As Block (2008b) notes, facilitation includes a wide range of activities. In most cases, the more radical and disruptive the new technology is, the greater the number of barriers the state is required to help surmount before successful markets can be established – classic examples include the railroad and automobile, both of which required the state to construct vast amounts of expensive infrastructures before they could mature. In less radical instances, facilitation may simply involve establishing standards capable of demonstrating to purchasers that the technology functions properly, or regulatory frameworks that legitimize the technology and allow investors and consumers to feel comfortable embracing it.

The facilitation effort in the climate/energy context has seen Washington establish more than 300 individual programs and laws intended to combat a series of 'barriers' to market entry. Barriers tend to be present at every point in the commercialization process, and can emanate from several sources including market failures (for example, monopolistic industry structures, externalized benefits and costs, misplaced incentives, and/or incomplete/imperfect information); government failures (for example, competing fiscal/regulatory policies that conflict with the promotion of new energy technologies); or miscellaneous difficulties related to the nature of technology markets (for example, lengthy learning curves, poor initial economies of scale, entrenched consumer preferences, weak supply chain infrastructure, lack of workers trained to work with new technologies, problems with intellectual property portfolios and so on), and can all be interrelated and mutually reinforcing with a range of other cultural and historical barriers (see Table 3.5).

In response, the federal government has established five broad types of initiatives, all tailored to the individual barrier. These include standards-setting (used to combat forms of market and technical risks); information dissemination (used to overcome imperfect information and lack of specialized knowledge); tax incentives (used to address high costs and external costs and benefits); state funded demonstration projects (used to overcome technical risks); and government procurement policies and contracts (used to address a lack of initial effective demand).

Curiously, however, in spite of nearly thirty-five years of fairly consistent effort coupled with hundreds of billions of dollars worth of

Table 3.5 Common barriers to novel energy technology deployment

Barrier Category	Sub-Barriers
Cost-Effectiveness	High costs Technical risks Market risks External benefits and costs Lack of specialized knowledge
Fiscal Barriers	Competing fiscal priorities Fiscal uncertainty
Regulatory Barriers	Competing regulatory priorities Regulatory uncertainty
Statutory Barriers	Competing statutory priorities Statutory uncertainty
Intellectual Property Barriers	Intellectual property transaction costs Anti-competitive patent practices Weak international patent protection University, industry, government perceptions
Other Barriers	Incomplete and imperfect information Infrastructure limitations Industry structure Misplaced incentives Policy uncertainty

Source: Committee on Climate Change Science and Technology Integration (2009).

investment and the deployment of a sophisticated DNS apparatus, the developmental state project around green energy innovation has been a failure. With the exception of a series of incremental (if nevertheless important) advances in a small range of technologies, Washington has come nowhere near the successes it achieved in supporting fields like IT, biotech, telecommunications, and others since the late 1970s. Indeed, the most 'radical' energy innovation generated by federal R&D over the past few decades has actually been the practice of hydraulic fracturing (commonly referred to as 'fracking'), which itself is only a modification of an earlier practice, and merely shifts a percentage of the US energy matrix to a slightly less-polluting form of fossil fuel usage.[10] The question, then, is why has this been the case? Why has the American developmental state continuously failed in the energy game? And what do these failures tell us about the limits of 'neoliberal statism'?

The failure of the DNS in the energy realm

There are at least two major problems with attempting to translate renewable energy into a developmental state project that have not been strongly present in previous state-led technology projects. The first problem is simply one of demand – more specifically, a lack of effective bottom-up demand for renewable energy. Put simply, the demand for 'clean' energy inputs in an economy is an *abstract* form of demand, given that, in the absence of a pricing mechanism, markets do not specifically seek or place value on 'clean' energy, but rather are generally inclined towards the least costly form of energy. The problem in this equation is, obviously, that the market for cheap energy has long been cornered by conventional fossil fuels, and the majority of renewable systems remain comparatively more expensive.

While this may seem obvious, this demand problem makes renewable energy an extreme outlier amongst developmental state projects – with past successful projects generally addressing only demands not already being met by cheaper established technologies. As a result, to date, the only renewables that have managed to become profitable and marginally competitive with conventional sources are very low-risk 'conventional alternative' technologies that benefit from very generous (yet very inconsistent) government production-subsidies and tax rebates – for example, biofuels derived from food crops, solar parks, and onshore wind farms (Yanosek, 2012; Rai et al., 2011).[11] As capital prefers to invest in projects that are lower risk and have a higher guarantee of profitability, a situation has materialized in which nearly seven-eighths of all renewable energy investment worldwide now goes to deploying this small number of technologies that would not otherwise be profitable without government subsidies.

The second problem, largely related to the first, is the extent of fossil energy's market entrenchment. Indeed, fossil fuel benefits not only from its price advantage over renewables, but also from the fact that its enabling infrastructures, global supply and value chains, consumer biases and behaviors and so on, are already well established and deeply entrenched, while those of renewable energy remain largely undeveloped (Yanosek, 2012).[12] Given the sheer volume of high-risk capital investment required to build these infrastructures and supply chains, few technologies have ever actually been afforded the opportunity to *become capable* of competing with fossil energy. This has been the case not only for large-scale energy production technologies (for example, coal carbon capture and sequestration systems and next generation nuclear reactors), but also storage, battery, and transport capacities that would allow renewable technologies and other forms of energy farming to fully mature.

Under normal circumstances, this lack of effective demand could be remedied by regulations on fossil fuel usage designed to artificially 'pull' renewable technologies to market. Such regulation would create politically-generated market demand for new technologies primarily by raising the price of fossil energy relative to renewables, thereby allowing them an opportunity to plausibly compete – and thus encouraging new market entrants and advancements in the technologies. The growth facilitated by these policies could, over time, help to establish a durable market capable of expanding of its own volition.

Ironically, however, the implementation of such policies has been precluded by the very politics of neoliberalism that the DNS was established to contend with. Indeed, the combination of the strength of anti-regulationist ideology in the US Congress (which has been a key cornerstone of contemporary American neoliberalism) along with the clientelistic power of fossil fuel interests has made the implementation of new federal environmental regulations all but impossible over the past three decades (for example, see Klyza and Sousa, 2008; Driesen, 2010; MacNeil, 2013). With regard to the former point, hammering on the trope that such regulation serves only to stifle the investment and innovation on which American economic growth and job security rely, proponents of this ideology have consistently managed to frame the regulation of fossil energy use as an attack on the income security of individuals, families, and small businesses, as well as an affront to the living standards that Americans have come to expect.[13] With regard to the latter, the disproportionate representation of coal-states (that is, states where coal extraction represents a significant element of the economy, and where substantial numbers of individuals are employed in the coal industry) in the US Senate has rendered the upper chamber something of a graveyard for climate legislation. Between 1999 and 2010, fifteen comprehensive and more than eighty non-comprehensive climate bills met their ultimate demise in the Senate, in spite of growing support for climate regulation from industry and civil society during this period (see Tables 3.6 and 3.7).[14] To make matters considerably worse, Republicans have consistently favored (and typically been successful in implementing) extremely generous tax cuts for fossil fuel corporations, thereby further enhancing the market position of conventional energy.

Table 3.6 Failed major climate bills in the US Congress, 2003–2010

Proposed Bill	Year	Intended GHG Reductions by 2020	Covered Sources	Primary Reduction Mechanism
Climate Stewardship Act	2003	Maintain at 2000 levels	Electric power; industrial; commercial; transportation petroleum	Cap and trade
Clean Air Planning Act	2004	15% below 2005 levels	Generators of electricity	Cap and trade
Climate Stewardship and Innovation Act	2005	15% below 2005 levels	Electric power; industrial; commercial; transportation petroleum	Cap and trade
Clean Power Act	2005	39% below 2009 levels	Generators of electricity	Cap and trade
Global Warming Pollution Reduction Act	2006	Maintain at 2010 levels	Electric power; industrial; commercial; transportation petroleum	Cap and trade
Low Carbon Economy Act	2007	Maintain at 2006 levels	Coal facilities; petroleum refineries; natural gas processors; manufacturers and importers of HFCs, PFCs, SF_6, and N_2O; aluminum smelters, and cement producers	Cap and trade
Climate Stewardship and American Innovation Act	2007	15% below 2005 levels	Electric power; industrial; commercial; transportation; petroleum	Cap and trade
Safe Climate Act	2007	15% below 2007 levels	Electric power; industrial; commercial; transportation petroleum	Cap and trade
Electric Utility Cap and Trade Act	2007	8% below 2007 levels (utilities only)	Electric utilities	Cap and trade for electric utilities only

Table 3.6 Failed major climate bills in the US Congress, 2003–2010 – *continued*

Proposed Bill	Year	Intended GHG Reductions by 2020	Covered Sources	Primary Reduction Mechanism
Climate Security Act	2007	19% below 2005 levels	Electric power; industrial; producers/importers of petroleum- or coal-based fuels; producers/importers of non-fuel chemicals	Cap and trade
Clean Air/Climate Change Act	2007	6% below 1990 levels (the US's Kyoto obligation)	Not specified	Not specified. Intended to add CO_2 emissions to list of regulated air pollutants under Clean Air Act
Global Warming Reduction Act	2007	15% below 2007 levels	Electric generation; motor vehicles; fuel	Performance standards with the option for an emissions cap and trade system
Clean Power Act	2007	17% below 2005 levels	Generators of electricity	Cap and trade
Clean Energy Jobs and American Power Act	2009	20% below 2005 levels	Electric power; industrial; producers/importers of petroleum- or coal-based fuels	Cap and trade
American Power Act	2010	17% below 2005 levels	Generators of electricity; petroleum fuels; distributors of natural gas	Cap and trade

Source: Center for Climate and Energy Solutions (2012).

Table 3.7 Selected failed non-comprehensive climate bills in the US Congress, 1999–2010

Proposed Bill	Year
S.1369: Clean Energy Act	1999
S.1949: Clean Power Plant and Modernization Act	1999
H.R.2569: Fair Energy Competition Act	1999
H.R.2645: Electricity Consumer, Worker and Environmental Protection Act	1999
H.R.2900: Clean Smokestacks Act	1999
H.R.2980: Clean Power Plant Act	1999
S.882: Energy and Climate Policy Act	1999
S.1776: Climate Change Energy Policy Response Act	2000
S.1777: Climate Change Tax Amendments of 1999	2000
S.1833: Energy Security Tax Act	2000
H.R.3384: Energy and Climate Policy Act	2000
S.556: The Clean Power Act	2001
S.1131: The Clean Power Plant and Modernization Act	2001
S.3135: The Clean Air Planning Act	2001
S.1333: The Renewable Energy and Energy Efficiency Investment Act	2001
S.1716: The Global Climate Change Act	2002
S.1781: The Emission Reductions Incentive Act	2002
S.1870: A bill to amend the Clean Air Act to establish an inventory, registry, and information system of U.S. greenhouse gas emissions	2002
H.R.3037: The Renewable Energy and Energy Efficiency Investment Act	2002
H.R.4611: National Greenhouse Gas Emissions Inventory Act	2002
S.389: The National Energy Security Act	2002
S.597: The Comprehensive and Balanced Energy Policy Act	2002
S.1008: The Climate Change Strategy and Technology Innovation Act	2002
S.1293: The Climate Change Tax Amendments	2002
S.1294: The Climate Change Risk Management Act	2002
S.932: The Conservation Security Act	2002
S.1255: The Carbon Sequestration and Reporting Act	2002
S.892: The Clean and Renewable Fuels Act	2002
S.17: The Global Climate Security Act	2003
S.194: The National Greenhouse Gas Emissions Inventory and Registry Act	2003
S.366: The Clean Power Act	2003
S.14: Energy and Climate Change Amendment to the Energy Policy Act	2003
H.R.1245: The National Greenhouse Gas Emissions Inventory Act	2003
S.2571: The BOLD Energy Act	2004
S.2984: Future Investment to Lessen Long-term Use of Petroleum Act	2004
S.3543: Ten-in-Ten Fuel Economy Act	2004
S.Amdt.815: Energy and Climate Change Act	2005
H.R.955: National Greenhouse Gas Emissions Inventory Act	2005
H.R.2828: New Apollo Energy Act	2005
H.R.5049: Keep America Competitive Global Warming Policy Act	2005
H.R.5642: Safe Climate Act	2006

Table 3.7 Selected failed non-comprehensive climate bills in the US Congress, 1999–2010 – *continued*

Proposed Bill	Year
H.R.6266: 21st Century Energy Independence Act	2006
H.R.6417: Climate Change Investment Act	2006
S.6: National Energy and Environmental Security Act	2007
S.133: American Fuels Act	2007
S. 1018: Global Climate Change Security Oversight Act	2007
S. 1059: Zero-Emissions Building Act	2007
S. 1073: Clean Fuels and Vehicles Act	2007
S. 1168: Clean Air/Climate Change Act	2007
S. 1324: National Low-Carbon Fuel Standard Act	2007
S. 1387: National Greenhouse Gas Registry Act	2007
S. 1411: Federal Government Greenhouse Gas Registry Act	2007
S.1874: Containing and Managing Climate Change Costs Efficiently Act	2008
S.1462: The American Clean Energy Leadership Act	2009
S.2877: The Carbon Limits and Energy for America's Renewal Act	2009
S. 2776: Clean Energy Act	2009
S.3464: Practical Energy and Climate Plan	2010

Source: Center for Climate and Energy Solutions (2012).

This has meant that, while the federal government *can* help to develop green technologies by funding and coordinating laboratory innovation through the DNS, it is largely powerless to create the market conditions necessary to support their success in the market.

Conclusion

To briefly recapitulate, the central argument of this chapter is basically three-fold. First, building on recent research on American innovation, I have noted that technological development in the US economy is considerably more state-driven than conventional literatures would tend to suggest. In particular, building on the work of Block (2008b), I have sought to underscore the influential role of a developmental network state apparatus present within the federal government since at least the early 1980s – an entity whose establishment was animated by policy-makers' attempts to foster accumulation and growth following the breakdown of the Fordist/Keynesian post-war order in the US. Secondly, I have argued that, over the past three decades (and rapidly accelerating over the past four years), a dedicated developmental state apparatus has emerged in the climate/energy realm, aimed at actively promoting the development and commercialization of green technologies. This system of R&D funding and networking mirrors the strategies used in past successful developmental state projects, with the federal government actively guiding the innovation process from the earliest stages of concept design, right through to demonstration and commercialization. However, the chapter's third primary aim has been to demonstrate that, in spite of the increasingly high levels of funding and qualitative assistance, the federal government's DNS project around climate/energy innovation has largely failed to produce any substantial results.

The failure of the DNS to successfully promote climate innovation underscores the primary limitation of a developmental state apparatus within a neoliberal state. The DNS emerged in the neoliberal era as a mechanism designed *only* to parlay innovation into economic growth and accumulation. It was never designed to use innovation as a means to achieve non-economic goals like environmental protection or emissions reductions. It was also designed *only* to develop technologies that fill market needs not already being adequately met by another technology (this could be said of past successful DNS projects around IT, biotech, telecommunications, and others). It is not designed or equipped to undermine existing technologies or industry participants already fulfilling the same abstract demand. To do so would require the state to actively regulate fossil fuel usage in order to undermine its dramatic price and infrastructural advantages in the current energy market. The politics of neoliberalism (and its hostility towards environmental regulation) mean that Washington does not have the capacity to do this. The result is that, while it can fund and network novel innovations throughout the laboratory and demonstration process, it cannot furnish the market conditions required to properly commercialize and mainstream these technologies. In short, while innovation has increasingly become the business of government in the US, successful state-led innovation in the climate realm will require capacities that neoliberal states do not possess – namely the capacity to overtly regulate massive swaths of the economy, and exercise immense political power to undermine an existing powerful industry.

This brings us back to one of the primary general themes about national innovation systems highlighted by Mikler and Harrison (2012). As the authors suggest, 'the institutions of national technology systems are primarily designed to achieve economic or political, not environmental, outcomes. Their primary purpose is to encourage domestic economic growth and industrial competitiveness with other nations' (Mikler and Harrison, 2012: 198). This is, indeed, the *sole* purpose of the American developmental state. Any positive environmental outcomes that may eventuate from climate innovation would merely be a happy coincidence for the DNS. Its singular objective is to establish American leads in high-tech sectors in order to generate greater domestic accumulation, profitability, and jobs, and revenues for the state.

What, then, would be required to render the DNS an effective mechanism for promoting climate innovation in the US? Two particular avenues seem somewhat possible. The first is a gradual shift away from the hostile politics of neoliberalism and anti-regulationism at the federal level, one that would allow the state to overtly challenge and undermine the reigning fossil fuel regime of accumulation, and develop a state-led innovation policy based explicitly on achieving environmental targets. Such conditions would be considerably more conducive to a DNS project in which CO_2 emissions reductions were the primary goal of climate innovation, and economic accumulation was merely an ancillary benefit. Failing this, a second option would be to have the mantle of climate and energy innovation taken up by a powerful institution within the federal government that is largely immune from the political impediments outlined above, and can engage in innovation for non-economic objectives. This option seems considerably more likely at present, with the Department of Defense (DOD) currently in the midst of establishing a vast manifest of renewable energy programs aimed at weaning the military off of traditional fossil fuels for defense-related imperatives. The military provides several key advantages in this scenario. First, with the DOD's R&D budget reaching as high as $80.92 billion in 2010, the military can fund innovative endeavors in a manner that would be largely unthinkable for other public and private organizations (Keller, 2012). Second, with defense spending accounting for 4.5 percent of GDP, the military's capacity to procure the technologies it develops can go a long way towards generating enough demand to guide a new technology through its earliest and costliest generations (Keller, 2012). Most importantly, however, the DOD's innovation agenda is not based exclusively on economic growth and accumulation. While most military technologies are eventually spun off into a civilian technology at some point down the road, the DOD is able to operate outside the narrow confines of the contemporary energy market.

Whatever the way forward may be, it seems likely that policymakers in Washington will be increasingly compelled to resolve this fundamental problem with state-led climate innovation. As other countries vigorously strive to dominate the multi-trillion dollar alternative energy market of the coming century, the state's interest in promoting renewed cycles of accumulation should be expected to increasingly trump the narrow interests of the 20th century fossil energy regime.

Notes

1 These efforts translated into the following wave of laws and initiatives: Stevenson-Wydler Technology Innovation Act; the Bayh-Dole Act; the Small Business Innovation Development Act; the National Cooperative Research Act; the establishment of a Program for Engineering Research Centers; the Federal Technology Transfer Act; the Advanced Technology Program; the Manufacturing Extension Program; the Defense Industrial and Technology Base Initiative; the High Performance Computing and National Research and Education Network Act; and the Small Business Research and Development Enhancement Act (see Block, 2008b).
2 O'Riain (2004) provides a helpful distinction between the DNS in operation in the US and parts of Western Europe and the Developmental Bureaucratic States (DBS) present in East-Asia in the post-war era. While the DBS model (epitomized by Japan's Ministry of International Trade and Industry) was a highly centralized bureaucratic entity designed to help domestic industry 'catch-up' with the technological advances of the West, the DNS, by contrast, as its name suggests, is effectively a networking effort that aims to help the nation's scientific and engineering communities navigate the most promising avenues for novel innovations that do not yet exist and commercialize them – a task for which the DBS model is obsolete since there are no existing competitors to ape. In contrast the centralized bureaucratic nature of the DBS, the work of the decentralized DNS takes place in thousands of labs, businesses, and universities across the country.
3 Full text of the Energy Policy Act of 1992 is available at: http://thomas.loc.gov/cgibin/query/z?c102:H.R.776.ENR
4 The point here is not to overstate the financial value of these investments. Put in context, they would represent only a small percentage of federal R&D, and certainly would hardly compare to innovation in other realms, like defence, for example. The point is rather to highlight the continuity of efforts by successive administrations and Congresses to begin providing foundational economic support for alternative energy initiatives.
5 These figures can be found at: http://www.whitehouse.gov/issues/energy-and-environment and http://www.energy.gov/recovery/
6 All defence and non-alternative energy related figures have been subtracted to get this amount in both the 2010 and 2011 budgets. Text of the FY 2010 budget and 2011 budget request can be found at: http://www.whitehouse.gov/omb/
7 These categories are borrowed (with modifications) from Block (2008b).
8 Examples include the United States Council for Automotive Research and its thirty component programs (of which the Partnership of a New Generation of Vehicles was among the most prominent).
9 There were three crucial pieces of legislation specifically tackling technology transfer. First, the Stevenson-Wydler Technology Innovation Act of 1980 officially added technology transfer to the mandate of all federally

funded research labs in an effort to guarantee the full use of the country's federal investment in R&D activities. The Act further provided for the ability to use federal funds for this purpose. Second, the Bayh-Dole Act (or the University and Small Business Patent Procedures Act) of 1980 gave federally funded small businesses the right to take full ownership of any intellectual property created in the course of their work in the labs. And finally, the Federal Technology Transfer Act of 1986 allowed the labs to establish Cooperative Research and Development Agreements (CRADAs) with private businesses. This arrangement gave the labs the option to share research costs with a private partner and negotiate an exclusive license to the resulting technology without the otherwise necessary public notice (freedom of information) requirement.

10 Worse still, fracking may help to drive down the price of fossil energy to a point where alternatives have even less chance of competing.

11 In the US, for example, tax credits and depreciation benefits make up more than half of the after tax returns of conventional wind farms and two-thirds of solar energy projects, while the federal government provides a subsidy of nearly $1.50 per gallon of corn-based ethanol (Victor and Yanosek, 2011).

12 This relationship is obviously dialectical to the extent that one of the main reasons why fossil energy is comparatively cheap is because of its entrenchment, and one of the reasons that it remains entrenched is its low price.

13 A good example of this framing came in the lead up to the House vote on the American Clean Energy and Security Act of 2009 – a bill that would have established a cap-and-trade system around greenhouse gas emissions for energy utilities and other selected sectors. Republicans and Democrats from coal states consistently cited a study prepared by the conservative policy think tank The Heritage Foundation which suggested that, over the long term, the legislation would 'reduce aggregate gross domestic product by $7.4 trillion; destroy 844,000 jobs on average, with peak years seeing unemployment rise by over 1,900,000 jobs; raise electricity rates 90 percent after adjusting for inflation; raise inflation-adjusted gasoline prices by 74 percent; raise residential natural gas prices by 55 percent; raise an average family's annual energy bill by $1,500; and increase inflation-adjusted federal debt by 29 percent, or $33,400 additional federal debt per person, again after adjusting for inflation' (Beach et al., 2009).

14 The most important element of this growing support has been the broad acceptance of emissions trading schemes by industry over the past decade – particularly following the formation of a pro-trading advocacy coalition called the United States Climate Action Partnership (USCAP) in 2007. Initially led by many of the same large energy conglomerates, banks, and mainstream environmental groups that first helped to put emissions trading on the global agenda in the 1990s (notable members include BP, Shell, Duke Energy, DuPont, General Electric, AIG, Lehman Brothers, General Motors, Chrysler, Dow Chemical, ConocoPhillips, Environmental Defense, Johnson & Johnson, World Resource Institute, Alcoa, and PepsiCo), USCAP has gradually consolidated a durable cross-sectoral alliance comprised of major firms representing virtually every major economic sector in the country (Meckling, 2011).

4
Colorado's New Energy Economy: Ecological Modernization, American-Style?[1]

Stratis Giannakouros and Dimitris Stevis

During his Administration (from January 2007 to January 2011) Colorado Governor Bill Ritter pursued a 'New Energy Economy' (NEE) strategy, one of the most ambitious and far reaching attempts at reorganizing the economy of a state in recent times. In this chapter we argue that the NEE was an attempt at ecological modernization (EM), rather than a set of ad hoc initiatives, because it sought to fuse innovation, economic and environmental goals through the political leadership of an activist state at the head of an alliance of environmentalists, industry and other societal forces. This strategy brought together and was shaped by sub-federal and federal forces and included and promoted a number of institutional innovations towards climate change, innovations that have been and continue to be the subject of serious political contestation both at the state and national levels.

A key question that emerges from our case is whether the NEE was an extension and manifestation of federal policy or whether this case underscores the increasing role of sub-federal governments in economic policy, in general, and climate policy, in particular (Aulisi et al., 2005; Byrne et al., 2007; Rabe, 2003). As we will argue, one cannot underestimate the presence of the National Renewable Energy Lab (NREL) and other federal organizations in Colorado. Nor can we bypass the impacts of the American Recovery and Reinvestment Act (ARRA) instituted in 2009 to confront the financial crisis. However, our case suggests that we need to pay attention to state-level dynamics *and* investigate the federal-sub-federal dynamic closely. This is supported by evidence of the increasing economic role of United States (US) states over the last thirty years.

We start with a discussion of EM and innovation, followed by a clarification of 'EM American Style' and the institutional reasons for it.

We then trace the NEE, with a focus on institutional innovations, starting with the adoption of the Renewable Energy Standard (RES) legislation in 2004 and continuing to its present prospects under a new state government and in light of the resurgence of natural gas extraction in the state and across the country. This trajectory offers a dynamic understanding of the NEE as an evolving and contested policy. We conclude by bringing together the broader lessons of this case with respect to sub-federal innovation for climate change, the impacts of federal-state relations in this process, and the prospects for EM in a liberal and federal capitalist system, like the US.

The lessons are firstly that institutional innovation is an important driver and enabler of climate innovation; and secondly that energy transitions are contested and uncertain processes prone to counter-revolution and reversal. Therefore, in thinking about the effectiveness of attempts at state-level environmental and economic reorganization historical path dependencies and institutional context matter greatly. Thirdly, sub-federal attempts at climate innovation are important and necessary, but insufficient in addressing the collective action problem climate innovation presents. Finally, the federal government plays an important role in both setting policy and providing resources to states, but would do well to pursue a more collaborative strategy with US states that supports existing state-level institutional innovations and policies.

Ecological modernization and innovation

The core assumption of EM is that the environmental degradation caused by industrialization can be 'solved' through innovations that decouple energy consumption from commodity production while simultaneously reducing waste (Hajer, 1995; Huber, 2000; Schlosberg and Rinfret, 2008; Jänicke and Lindemann, 2010). In this regard, EM can be seen as a pragmatic theory seeking to balance the competing claims of environmentalists and industry:

> [EM] did not develop primarily from a pre-existing body of social-theoretical thought...Instead, ecological modernization thought has been more strongly driven by extra-theoretical challenges and concerns (e.g. about how to respond to radical environmentalism and how to conceptualize eco-efficiency improvements that are currently linked to new management practices and technical-spatial restructuring of production), Ecological modernization has

essentially been an environmental science and policy concept (Buttel, 2000: 64).

There are vibrant discussions as to which policies fit within the parameters of EM (Deutz, 2009), whether EM can be differentiated into weaker and stronger versions (Christoff, 1996) and whether even the strongest versions of EM are transformative (Warner, 2010). We are not going to explore these debates here, but we do feel that we need to clarify up front the main categories of EM and the significance of innovation for each. We start with 'strong' EM because we do not think it is applicable in the US or most cases that may be considered instances of EM. We then proceed to 'weak' EM and Industrial Ecology that are concepts that may denote the NEE. In differentiating between 'strong' and 'weak' versions of EM it is important to consider both the environmental depth of reforms and the scope of those reforms simultaneously. Instances where reforms of great depth and broad scope are present are classified as 'strong' EM. Whereas, instances where reforms are environmentally superficial, parochial in their scope, or both are classified as 'weak' EM.

The stronger versions of EM seek to reorganize the whole economy and are particularly aware of the possibility of the displacement of environmental harm due to national strategies that are impervious to transnational impacts (Hajer, 1995; Christoff, 1996; Toke, 2011). In this context, institutional innovations aim at transforming the decision-making calculus of producers and consumers, including the extinction of particular activities and sectors. However, it is important to note that critics such as Rosalind Warner (2010) argue that even strong EM, although certainly more profound than other types of EM, is still not transformative enough in terms of its ecological priorities.

What has been termed 'weak' EM covers a lot of ground between more and less significant *reforms* at the level of whole sectors and even the whole political economy, but largely within the existing economic paradigm (Gouldson and Murphy, 1997; Mol and Sonnenfeld, 2000; Young, 2000). It does share with stronger EM an understanding that institutional innovations are necessary. In this respect, it goes beyond industrial ecology, the next category, in that it does aim at whole sectors or the whole economy and does recognize the need for broader institutional reform rather than simple eco-efficiency.

The weakest approach to EM is industrial ecology (IE), which tends to privilege technical and managerial innovations (Frosch and Gallopoulos, 1989; World Business Council for Sustainable Development, 2012;

Ayres and Ayres, 2002; Lifset and Graedel, 2002). Such innovations take place within the parameters of individual economic units, largely at the level of corporations and organizations.

Despite any differences between 'strong' and 'weak' approaches to EM, their interpretation of innovation is an institutional one. Far from being a side product of technical innovations, EM is driven by political choices about the organization of the political economy. According to Lam et al. (2005: 107) institutional innovation involves, 'new norms and behaviours which private or public institutions adopt to stimulate technological, social and institutional environmental innovation during processes of ecological restructuring'. Such an approach is open to the possible forces behind institutional innovation, including the role played by NGO's and citizens groups which seek to reconfigure existing institutions and policies around social objectives, in this case low carbon objectives (see Wüstenhagen et al., 2007; Bergman et al., 2010; Seyfang and Smith, 2007).

At the risk of confusing the reader here we find the distinction between 'weak' and 'strong' innovation offered by Jänicke and Lindemann (2010) to be useful in differentiating among different types of weak EM, as well as evaluating IE (strong EM is only consistent with strong innovations). According to them, it is possible and necessary to distinguish between different types of innovation in terms of their ecological effectiveness. Strong innovations cannot be simply smart but must also have broad and deep environmental impacts. For example, efficiency gains at the level of the unit would not be ecologically effective because they are likely to be cancelled by the impacts of more aggregate consumption and production – what is often referred to as a rebound effect (Greening et al., 2000; Sorrell, 2009).

In evaluating the institutional innovations for climate change associated with the NEE period, therefore, we can distinguish between stronger innovations, that can actually protect the climate, and weaker innovations that have marginal or even negative impacts. In either case, to the extent climate innovation is currently taking place in the US, it is within the parameters of weak EM, on one hand, and IE on the other.

Ecological modernization and political authority

In addition to their common emphasis on innovation, EM perspectives also share an emphasis on the role of political authority. EM does not perceive the state to be in direct conflict with the interests of industry, but rather as a collaborator in a process towards a more efficient use of resources and decreased pollution (Jänicke, 1991; Weale, 1992; Barry and Patterson, 2004; Barry, 2007; Jänicke and Lindemann, 2010). This is, in fact, one of the reasons why a number of EM analysts have doubts as to whether industrial ecology is a type of EM. However, we feel comfortable in saying that weak and strong EM share a state-centered approach to environmental innovation and that IE, tacitly if not rhetorically, welcomes and often requires public and regulatory support. This is evidenced in the way tax breaks, incentives and the protection of intellectual and private property rights are manifested at the municipal, state and federal-level with regard to economic development schemes and business support.

The centrality of an activist state automatically raises doubts as to whether there can ever be EM in the US. Hunold and Dryzek (2005: 14) have argued that the explicit adoption of EM has also been resisted largely as a result of an adversarial culture in the US wherein policy discourse is, 'stuck in an old-fashioned standoff between supporters and opponents of the environmental policy regime established around 1970, and barely updated since'. While business interests have succeeded in framing environmental policymaking as a trade-off with economic growth, successive US administrations have done little to counter this framing in an explicit fashion.

A more persuasive explanation, in our view, is the same one behind any efforts at comprehensive economic policy, characteristic of liberal capitalism (Hall and Soskice, 2001a; Morgan and Kristensen, 2006; Block, 2011). It is not the case that capital does not engage or even promote the role of the state under this system. However, it promotes a state that enables rather than leads. The state, in this view, may enable firms to act on the basis of markets signals but can also shape the institutional make up of the markets themselves. The tensions between forces that want a more activist state, whether for social, military or economic reasons, and those that want a more enabling state has resulted in what Mettler (2011) calls the 'submerged state'[2] and, we may add, a very unevenly submerged one. As she states (Mettler, 2011: 804):

> Especially during the past two decades, the submerged state has nurtured particular sectors of the market economy and they have in turn invested in strengthening their political capacity for the sake of preserving existing arrangements. As a result, the alteration of such arrangements has required either defeating entrenched interests –

which has proven impossible in most cases – or, more typically, negotiating with and accommodating them.

This dynamic is also evident with respect to the environment. During the 1970s and, again, during the last decade, there have been various federal-level efforts to promote a greening of the economy at the national level. Although terms such as 'industrial policy' are avoided for political reasons – that is, a national aversion for coordinated economic policies – and EM because of unfamiliarity in the US, there is evidence of federal-level initiatives with significant impacts on innovation. Behind these efforts are important social alliances, mostly amongst Democrats, who have called for broad national programs to green the economy (Block, 2008a; Pollin et al., 2008).

As Block and Keller (2011) and their collaborators demonstrate, this centralized, if not well coordinated strategy, of the federal government has had far reaching impacts upon various US states and the public. As the physical endpoints of laboratories and investment, a number of states have been profoundly affected by federal initiatives. For example, Andrew Schrank (2011) points to the state of New Mexico as the beneficiary of targeted federal investment through the levers of the federal government's centralized industrial policy. Using Los Alamos and its eventual offshoot, Sandia Laboratory, as a tool of technology driven policy, the federal government created an environment that would lead to an embrace of technological innovation and commercialization of renewable energy technologies in what would otherwise be an 'oil patch' state. Schrank (2011) argues that it is precisely the government's strategic intervention through Sandia laboratories that has recast the economic identity of New Mexico and spawned the entrepreneurial and political activity that currently embraces the renewables sector within the state.

Sub-federal ecological modernization

The origins of sub-federal industrial policy date back to the 1980s when, in response to the absence of an industrial policy at the federal level, many US states adopted an interventionist stance with regard to local and regional economic policy. There were two sets of motivations behind this activism. First, old industrial states started feeling the impacts of neoliberal globalization and outsourcing, and the Republican unwillingness to protect them (Rupert, 1997). Second, Southern states saw an opportunity to attract industrial investment, whether from other US states or abroad (Cobb and Stueck, 2005).

According to Susan Hansen (1989), this has led to considerable changes in both the substance of state economic policies and in the process through which they are developed, implemented, and evaluated. In the US context, this industrial policy that occurs at the state-level assumes a meso-corporatist hue with an emphasis on competitiveness against other states and other countries (Silver, 1987; Silver and Burton, 1986; Hayden et al., 1985). Peter Eisinger (1990) has argued similarly that state-level economic policy takes a programmatic form such that its defining elements are common to most national industrial policies.

Over time, as the novelty of state-level industrial policies has given way to an acceptance of the role of states in filling the gaps in federal-level industrial policy, state-level economic development strategies have become increasingly more sophisticated and debate over the efficacy of various approaches has proliferated. For the most part, these arguments have used the language of 'waves of economic development' looking at the role of incentives, leadership, information, and brokering in state economic development (see Clarke and Gaile, 1992; Eisinger, 1995; Bradshaw and Blakely, 1999; Hart, 2008 and Taylor, 2012).

More recently, scholars have also begun to look more closely at the nexus between energy and economic development at the state-level. Energy-based economic development has received a significant boost from ARRA and Department of Energy (DOE) loan grants since 2009 (see Carley et al., 2010; Carley et al., 2012). Other examples of New Energy Economy/Green Economy state-level initiatives, in addition to Colorado's NEE have also begun to emerge, such as in North Carolina (Debbage and Jacob, 2011) and Rhode Island (Rhode Island Economic Development Corporation, 2010).

In a growing number of instances, environmental voices appear to be playing an important role in driving economic development and industrial policies at the sub-federal level. In many respects, the participation of environmental interests in policymaking bodies at this level appears to be an American style EM. For instance, Pellow et al. (2000) and Scheinberg (2003) have argued that the modernization of waste management practices in North America constitutes an example of EM in practice. Scheinberg (2003: 72) in particular, contends that this evidence is not located at the level of the nation state, but rather is, '"down" from the level of the nation state to that of state, province, county and municipality'. In her estimation, 'it is the aggregation of all the small and medium-scale efforts that gives the picture of ecological modernization North American Style...in the North American context it is not industry-nation state relationships that are changed, but the relationships between industry and government, which, in US context, means state, county and local government'.

Schlosberg and Rinfret (2008) argue that a perceived shift is occurring as EM-influenced discourse is becoming more mainstream, promoting efficiency, developing new technologies and defining economic growth and environmental quality as a win-win scenario. However, given the federal government's reluctance to incorporate EM into federal policy, the task of implementing EM policy has fallen largely to private companies and individual states (Schlosberg and Rinfret, 2008: 259). Additionally, they observe that in most cases where industry has worked with the federal government to create new legislation, it has not taken the form of a more inclusive corporatist arrangement where environmental voices have a seat at the table. This has resulted in the adoption of EM policies in a less coherent and decentralized manner that in effect lacks the efficacy of more centralized European corporatist approaches (Gibbs, 2000). In practice, this state-level adoption of EM has manifested itself in the form of policies related to carbon emissions and renewable portfolio standards as well as more integrated approaches by state governments that promote 'clean energy economies' that address economic, environmental and climate concerns simultaneously, as in the case of Colorado's NEE.

Colorado's new energy economy

Colorado has long been the recipient of federal investments, largely military but, also, research related, such as the Institute of Arctic and Alpine Research (INSTAAR) (1951), the National Center for Atmospheric Research (NCAR) (1960) and the Joint Institute for Laboratory Astrophysics (JILA) (1962). In 1974, with the establishment of the Solar Energy Research Institute (SERI), the state entered the network of federal laboratories and cemented its position as a key player in alternative energy research. During the Carter Administration it was the recipient of a large budget for research into solar energy and the popularization of existing technologies, such as passive solar energy. The budget was cut by 90 percent during the Reagan administration but the institute survived and was renamed the National Renewable Energy Laboratory (NREL) in 1991. Since its inception, NREL has served as a conduit to federal research, supporting statewide activity in alternative energy. Although the budget for NREL has fluctuated greatly over the past several decades, dependent upon the political climate in Washington, its presence has fostered a great deal of activity around energy technology and innovation in the state of Colorado similar to the role of Sandia Labs in New Mexico, or Lawrence Berkeley and Bell Labs have in the Bay Area.

The key role of NREL may suggest that developments in Colorado were nothing more than an outcome of federal-level policy. However, state-level factors have also played an important and often leading role in the state. For example, in addition to NREL's presence in the state, organizations such as the Rocky Mountain Institute (RMI), founded by Amory Lovins in 1982, have promoted awareness of energy issues across the state of Colorado. For many the RMI is the *par excellence* advocate of solving environmental problems through innovation in the US, perhaps leaning in the direction of IE (Hawken and Lovins, 1999). Over the years, the RMI has been joined by various environmental and energy advocacy organizations (Environment Colorado, Western Resource Advocates, Colorado Renewable Energy Society (CRES), Colorado Solar Energy Association (COSEIA)) with a focus on fusing economy and ecology. These organizations have played a key role in 'localizing' the drive towards renewables.

Many of these developments have been supported and often facilitated by a culture that places value on environmental preservation, allowing for strategic cooperation between naturalists and 'environmental innovators' (though not always – see Cohen, 2006). To this end, the Colorado Constitution and its amendments are a useful register of Colorado's environmental attitudes. For example, the state's anti-nuclear amendment prevents any testing in the state without a citizen referendum. Additionally, environmental coalitions have played an important role in state economic development. Denver's successful bid for the 1976 Winter Olympics was derailed largely over environmental concerns. More recently, this history of environmental activism has shown its continuity with the present as a new environmental coalition has emerged to contest plans to explore a bid for the 2022 Olympics on the grounds that the exploratory committee does not have sufficient environmental representation (Williams, 2012). It can be argued that over time, the environmental forces present within the state have managed to create path dependencies that have simultaneously advanced and reinforced the state's environmental temperament.

Yet, it is well known that Colorado has also been an extractive economy since Europeans first consolidated their control during the mid-nineteenth century. For most of the state's history the extractive industries have successfully curtailed environmental interests and have dominated Colorado politics. To provide some context for the politics of the NEE, in 1997 the State of Colorado contracted Recom Applied Solutions to draft a GHG emission inventory and policy action plan. The Department of Public Health and Environment, in particular, hoped to use recommendations emerging from the analysis that would provide a blueprint for a strategy for air pollution within the context of earlier state programs targeting energy efficiency and conservation. However, the bold recommendations contained in the report infuriated the coal and mining industries as well as the state's utilities, causing an intensive backlash, ultimately damaging further prospects for GHG regulation at that time (Rabe, 2004). Likewise, in 1998, the Colorado legislature passed Joint Resolution 98-023 with strong bipartisan support that pointed to the economic costs of GHG regulation and implored President Clinton not to sign the Kyoto protocol. Language in the resolution further discouraged state or federal agencies from taking action to reduce greenhouse gases (JR 98-023, 1998). These examples of the contested nature of Colorado climate innovation serve to put later achievements, in the form of the statewide RES and subsequent legislation undertaken by the Ritter administration, in context.

Laying the foundation (2004–2007): The Renewable Energy Standard as institutional innovation

The origins of Colorado's NEE can be traced back to several key developments beginning in the 1990s and continuing into the 2000s. During this time period, advances in wind technology had lowered the unit cost of wind energy making it more competitive with fossil fuel generated energy. As noted, rising public concern *within the state* of Colorado over carbon emissions, environmental protection and pollution led to increased pressure on utilities to increase deployment of renewable energy alternatives to coal. This led to the creation of the Windsource program in 1998 by Xcel energy, Colorado's leading utility. At the same time, the Public Utilities Commission designated wind as a 'least cost' resource further pushing Xcel to build a wind energy infrastructure. These developments would embolden environmental advocates to push for a statewide RES and challenge the extractive energy alliance noted in the previous part.

Amendment 37 was a successful ballot initiative that established a statewide RES requiring investor-owned utilities (after strenuous lobbying, rural electric authorities were exempted from the new standard) to produce 10 percent of their electricity from renewable sources without exceeding a 2 percent ratepayer increase cap. The ballot initiative followed three failed attempts to pass an RES in the state legislature and was the first voter-approved RES in the country (Rabe, 2008). The use of direct democracy has been a policy tool employed by many US states over the past century, more recently used in controversial arenas such as environmental and energy policy (Guber, 2003). Although nearly half of all US states allow citizens to make laws directly through gathering signatures on petitions and then having statewide votes, every Western state with the exception of New Mexico allows for this brand of direct democracy. The near universal presence of ballot initiatives and referendums in the West is a product of the confluence between the development of many Western state constitutions in the early 1900s, an era of populism, progressive movements and citizen concern over the corporate influence of railroads, banks, mining and steel upon nascent legislatures. More recently, environmentalists have deployed these tools of direct democracy on many fronts, ranging from conservation efforts in Wyoming (Shanahan, 2010) to the municipalization of Boulder Colorado's electrical utility as a means of promoting renewable energy in the municipality over the longer term (Cardwell, 2013).

The passage of Amendment 37 is the cornerstone of the NEE for two reasons. First, the RES can be considered to be a key institutional innovation that would drive subsequent attempts to both foster and attract dynamic technical innovations in Colorado. Second, the passage of the Amendment deepened, sustained, and emboldened the political forces behind it. However, it is important to note here that the political forces behind Amendment 37 came from two different worlds. The majority were environmentalists and renewable industry advocates who valued environmental goals. But significant support also came from anti-environmentalists and climate skeptics whose primary interest was in new sources of income for their eastern Colorado constituencies, where the wind farms were expected to be built.

Adding to the Amendment 37 momentum, the Energy Policy Act of 2005 also contributed to the development of renewable energy by providing tax incentives and loan guarantees. Changes at NREL also helped. While it mainly provided federal support for conventional energy development, NREL's annual funding jumped from $209.6 million in 2006 to

$378.4 in 2007 (and $536.5 in 2010). It is also worth noting here that in January 2005 Dan Arvizu became its Director and Chief Executive. Arvizu had played a leading role in turning Sandia laboratories into an incubator (see Schrank, 2011; NREL, 2012). NREL was also a major force behind the Colorado Renewable Energy Collaboratory (CREC), which started operations in 2006. The CREC tied together NREL and Colorado's three major universities: University of Colorado-Boulder, Colorado State and Colorado School of Mines. In addition to the CREC, the Rocky Mountain Innosphere in Fort Collins and the Innovation Center of the Rockies in Boulder, both DOE Clean Energy Alliance Partners, have played an important role in encouraging indigenous clean-tech startup activity along the I-25 corridor (DOE, 2012a). Here, we clearly see state-level forces articulating themselves around federal policy and benefitting from it.

Bill Ritter filed papers to run for Governor of Colorado in May 2005, five months after leaving Denver's District Attorney's office. Early on Ritter chose to outline a vision for Colorado's future that would differentiate him from his opponent. One of the issues that Ritter began to explore was the potential for Colorado to develop a clean energy economy and diversify the state's energy portfolio. He met with representatives in the wind and solar industry and decided to include clean energy under his campaign umbrella of the 'Colorado Promise' (Ritter, 2006). For Ritter, the NEE represented a long-term energy strategy for Colorado. Importantly, Ritter also made a link between clean energy and job creation, thereby connecting the energy, environmental and economic benefits of the NEE – hence New Energy *Economy*.

Governor Ritter described the NEE as an ecosystem, guided by four principles: diversifying energy, protecting the environment, promoting economic development, and promoting equity. In his view, it was not possible to promote economic development of the kind that the NEE envisioned without appropriate policy, cutting edge technology, adequate financing and a workforce that is appropriately trained. Over the years, the various actors in support of the NEE have tended to emphasize one or some combination of these dimensions or frames. In some cases these choices were antagonistic, while in others synergistic.

With Ritter's adoption of the NEE, then, we see the outlines of a comprehensive strategy that aimed at reforming the interface of economy and environment at the level of the whole state and doing so by promoting integrated policy innovations. At the heart of the strategy was an alliance of political leaders, environmentalists, and people from the renewables industry, which was successful in winning the governorship as well as both houses of the legislature in the Fall of 2006. It is not unreasonable to suggest that Ritter's strategy fits well within the parameters of EM.

Towards EM: Innovations outside and inside the state (2007–2008)

The first two years of the Ritter Administration produced a range of initiatives both external and internal to the state apparatus, helped a great deal by the Democratic Legislature. In 2007, the increase in the RES from 10 to 20 percent, while maintaining the 2 percent rate increase cap, was less contentious because the major utility, Xcel Energy, had realized that it could adjust to these standards with relative ease. Moreover, it saw wind power as an opportunity to hedge its future, especially as Environmental Protection Energy (EPA) rules were targeting its coal-fired plants. That Xcel was comfortable using wind energy to help reach that RES is indicative of a certain degree of innovation-forcing.

Conversely, rural electric authorities and municipal authorities, with some exceptions, were as opposed to the new RES as they had been to the earlier one. Thus, they were expected to achieve a lower, 10 percent RES. The reasons for that opposition are not relevant here. We are noting the exemption, however, because they account for about 40 percent of the state's energy production and consumption.

Given the relative ease with which the increased RES was passed, the major challenges had to do with the organization of the state itself. Early on Ritter realized that he could not push forward with the NEE without internal change. One such change upgraded the state's Energy Office, which had been part of the Office of Economic Development Department. The newly named Governor's Energy Office (GEO) developed into a key facilitator in the development and adoption of a collection of policies and programs associated with the NEE.

A second major organizational change was the opportunity to appoint the new members of the Public Utilities Commission shifting its balance towards environmental and consumer interests. The third, and most contentious change, was the reorganization and expansion of the Colorado Oil and Gas Conservation Commission (COGCC), previously captured by industry, to accommodate a more diverse group of members with concerns about wildlife, public health, and the environment. As we were told by a person with close knowledge of the developments, 'the first big fight with the oil and gas industry wasn't the 20 percent renewable energy standard bill. The first big fight was the change in the make up' of the COGCC. The continuing significance of this reform lies in the current efforts to undo it and return the Commission to its pre 2007 composition. Despite the reform of the Commission it remains woefully inadequate to the task of overseeing the industry. For example, there are only sixteen full-time inspectors for about 50,000 wells (Presentation by Tom Kerr, acting director COGCC, Fort Collins, 23 February, 2013).

Changing the rules for drilling, particularly hydraulic fracturing with horizontal drilling, was the main reason for reorganizing the Commission. Over the years, landowners had fought against drilling (and other forms of extraction) because the law establishing split estate[3] allows drilling under one's surface property. Horizontal fracturing has made this problem even more vexing. Writing rules that would protect surface property owners and the environment proved to be extremely difficult as the gas industry claimed that the new regulations would be 'job killers', forcing companies to leave Colorado's gas fields. During 2008, the industry attacked the original iterations of these rules through a statewide campaign in which it enrolled local supporters. After an extremely long and contentious process, rules for locating and managing drilling were weakened.

There is persuasive evidence that the strategy of the Ritter Administration during 2007–2008 can very well be considered one of EM. The increase in the RES could in fact diminish or slowdown GHG emissions, environmental concerns were well integrated into decision-making processes and the role of political authority was front and center in this process. However, we need to consider these developments as taking place within some important limitations that may lead a number of analysts to consider the innovations weaker than they seem, in the language of Jänicke and Lindemann (2010).

Despite these considerable institutional innovations Ritter was not able to pass a comprehensive and explicit climate policy. The reassertion of the extractive face of Colorado, in the form of the aggressive opposition of the gas industry to the strongest versions of the drilling rules, ensured that The Colorado Climate Action Plan, issued near the end of Governor Ritter's first year in office in 2007, was more aspirational, softly pushing for climate protection policies. In addition to hostility from coal and gas, existing cleavages within the coalition that had supported the earlier RES standards rendered it ineffective in pushing for stronger climate action. As we have noted, that coalition included people who were anti-environmentalist and narrowly interested in channeling economic development to depressed rural areas that have been strong supporters of the Republican Party. For this reason, over time, environmentalists in the state have had to strike certain bargains that would allow them to pursue environmental goals as a collateral benefit of economic development and energy diversification.

So what were the outcomes? We have noted that EM seeks to integrate environment and economy and that the NEE was envisioned as a strategy to reshape the economy of Colorado. In this regard, the pursuit of green manufacturing and related R&D was central to the Ritter Administration. The Office of Economic Development and International Trade (OEDIT), along with GEO, worked intensely to attract the Danish company Vestas, the world's largest wind turbine manufacturer, to Colorado. Negotiations, that had started during the previous administration attracted Vestas to the state because of its well-educated work force, its railroads, an underutilized steel mill in the city of Pueblo and the wind potential of the region. As we have been told, the factors that led Vestas to choose Colorado over a number of other competitors who offered better financial incentives were the market certainty provided by the RES, and the overall strategy of the Ritter Administration. However, it is important to note here that incentives did and continue to play a role in economic development in the state. For example in 2011 alone, renewable energy companies in Colorado received $2.4 million in incentives from OEDIT (Jorgensen, 2012).

Vestas' move to Colorado is an important ingredient in understanding the NEE and its prospects. Clearly, wind power installation will facilitate reaching higher RES in the state and beyond. Most importantly, the presence of such a major company in Colorado shifts the balance between extraction and green manufacturing in the state. From 2007–2010, Vestas committed themselves to an investment of over $1 billion in four plants, generating 2,000 new jobs. This bold move, along with the NEE policies and NREL's increased activity, attracted other companies to establish a home base in Colorado, such as the German solar-system company Wirsol, and suppliers such as Bach Composite (Hartman, 2011; Zaffos, 2011). At the same time, companies already operating in Colorado increased their capacity. For example, Woodward Governor responded by expanding production in wind turbine inverters as soon as news of the Vestas agreement was made public. In addition to manufacturing facilities, a number of companies such as Siemens, Vestas, and Dragon Wind also opened R&D facilities to take advantage of emerging networks and economics of agglomeration.

This combination of R&D and manufacturing is worth noting because there are many instances throughout the world where manufacturing is not accompanied by R&D. Conversely, there are many instances in the core of the world economy where R&D is not accompanied by manufacturing. Colorado's approach sought to promote both in close proximity with the intention of creating a climate of knowledge spillover, agglomeration and green regional clustering (see Lewis and Wiser, 2007; Cooke, 2010). However, there are many other places around the country that have larger workforces and markets that compete directly for the same kinds of industry that Colorado wants, for example Michigan, Indiana, Ohio and other 'rustbelt' states. Colorado's main competitors include the intermountain states as well as Oregon/Washington and Iowa/Minnesota. In light of the competitive advantages of various states, it is doubtful that Colorado would exhibit the same strength in manufacturing and research and development in the renewable energy sector without the policies of the NEE.

Weakening EM: Extraction and crisis (2009–2011)

While the political and administrative reforms of the first period were undertaken by a strong Administration heading a strong economy, the second period (2009–2010) is colored by the financial crisis and the rise of natural gas as a key factor in state and national energy policy. In addition, as a result of the 2008 elections the Ritter Administration lost the State House and thus the significant advantage that it had from 2007–2008. The institutional innovations that took place during that second period have to be placed within these dynamics.

During 2009–2010 as the national crisis was unfolding and the EPA was threatening to shut down existing coal-fired plants, the Ritter Administration was considering the creation of three bills: one for an increase in the RES, one for certification of the workers in the industry, and one for green jobs training. In 2010, after a great deal of internal debate and negotiations involving environmentalists, unions, and industry (including gas but not coal), the RES was increased to 30 percent by 2020 for investor owned utilities (Hartman, 2011; Ritter, 2010; Zaffos, 2011). This bill, which included provisions for certification that were acceptable to labor unions, passed by a much smaller margin than the 2007 increase to 20 percent. However, it helped keep the coalition together, now with the acquiescence of the gas industry. This was necessary in order to pass the last major policy of the Ritter Administration, the Clean Air Clean Jobs Act (CACJ).

The main components of CACJ were to increase the market for Colorado's own natural gas and to avoid stricter regulations from the United States EPA. Natural gas was appealing because it could be integrated relatively easily with wind energy. As we noted above, one of the major innovations of the first period was the reorganization of the COGCC with the intention of writing drilling rules that protected surface property owners and the environment. During 2008 and 2009, the gas industry unleashed a statewide campaign against the first set of new rules, leading to their modification. More importantly, their offensive led Ritter to declare during 2009 that gas was now 'mission critical' and not simply a bridge fuel. From the point of view of the Ritter Administration, the unfolding economic crisis added impetus to that rapprochement as increased gas production promised immediate economic benefits. In addition, it allowed a solution to the pressure emanating from the EPA and involving Xcel's coal-fired plants. For the gas industry, moving some of these plants to gas was a desirable long-term commitment while Xcel secured long-term prices. The details and dynamics of these arrangements are worth another paper but the point should be clear that gas extraction was now in the tent.

In addition to state-level politics, the national financial crisis and national politics (in addition to the EPA) played a very important role during this period. Despite the crisis, the renewables industry, particularly wind power, was moving along but was not expanding as fast as expected earlier. Falling gas prices, along with uncertainty over renewal of the production tax credit for wind, started taking their toll in the form of fewer contracts and, thus, decreased manufacturing output. ARRA sought to provide a stimulus to the renewable energy sector but, as we will discuss shortly, it did not do so in a manner that would allow the NEE itself to grow under these conditions of crisis.

As the economic realities of the financial crisis began to be felt across the US in 2009, changing the political environment at both the state and federal-level, the Ritter Administration was forced to moderate its support for the environmental components of the NEE, leaning more in the direction of job creation and economic development based on extraction. The drilling rules, which the Administration considered a model for the country, provided a layer of comfort. Here, therefore, we have an institutional innovation that promoted or at least legitimated the proliferation of horizontal drilling with fracturing. The issue remains highly contested with at least two municipalities currently challenging state authority on the subject.

The ARRA affected Colorado and the NEE in a variety of ways. Of the approximately $5.5 billion approved for Colorado about $1.4 billion came from the DOE and hundreds of millions more from the NSF and NASA. Less came from the Department of Labor and the EPA. A great deal of that money went to federal projects and organizations based in Colorado but operating around the world. Thus, they did not contribute substantially to the NEE. The Alliance for Sustainable Energy, the operator of NREL, received close to $300 million. However, much of the support that goes to NREL is also for national programs and although there was an impact on the NEE, it was mostly indirect. In addition, ARRA extended loans of about $400 million to a local company, Abound Solar, and about $90 million to a local solar project (Cogentrix) (DOE, 2012b). While that money did not support the NEE strategy directly one can argue that it did so indirectly through local employment and research, especially since Abound Solar was the result of research at Colorado State University. Thus, the subsequent failure of Abound has had negative implications both for the DOE and the local development of cutting edge solar technologies and manufacturing.

Finally, ARRA provided the state with about $140 million to be spent on energy conservation and efficiency efforts, most of which was sent through the Governor's Energy Office (Zaffos, 2011). While some of that money was used to promote innovation at the point of the consumer, there was not enough time nor funds for significant innovations.

On balance, then, ARRA reaffirmed federal commitments and priorities, made choices with respect to technologies (Abound and Cogentrix) and helped the state with resources that allowed it to weather the financial crisis. But, as previously mentioned, the uncoordinated nature of the federal intervention did little to buttress the institutional changes brought about by the NEE at the state-level. Rather than aligning stimulus money with existing state policies, for example through block grants, the federal government chose to target strategic national *economic* priorities.

Undoing EM? Digging up the 'old' energy economy

Since January 2011, when Governor Hickenlooper took office, the NEE strategy has been on the defensive for reasons already apparent during the previous period. The spectacular growth of natural gas extraction has led to a hardening of positions between environmentalists and supporters of gas. Early on in his administration Governor Hickenlooper made it clear that he intended to diversify the

Governor's Energy Office, ensuring that conventional forms of energy were better represented – a goal he has largely achieved. As a result, members of the previous administration are concerned that the GEO will be completely changed to serve as a means of promoting fossil fuel energy in Colorado, particularly natural gas (Lynn, 2012). Hickenlooper has attempted to make changes to the COGCC and has made clear that he will litigate on behalf of natural gas interests where communities resist horizontal fracturing (Berwyn, 2012; Rascalli, 2011). The CREC is now just the Energy Collaboratory. While Hickenlooper is not opposed to the NEE in principle, it is unclear whether he will adopt many of the environmental elements of the NEE for his own administration.

So far, some of the innovations of the NEE period are holding. These include the RES and the composition of the Oil and Gas Commission. In this regard, we see evidence of continued inertia with regard to the RES. On April 30, 2013 the Colorado House passed legislation, which doubles the RES requirement for rural electricity cooperatives from 10 percent to 20 percent with a 2 percent rate cap (Moylan, 2013). However, to the degree that the state moves towards extraction, these stronger institutional innovations will become weaker. Ambitious RES standards become less impressive if overall fossil energy consumption increases. And the above-average methane leakage in the form of fugitive emissions from horizontal fracturing more than negates any GHG savings from moving away from coal.

In addition to the bankruptcy of Aboud Solar and the retrenching of GE's solar production plans, the future of Vestas, and thus green manufacturing, is also in doubt. The company has already closed its North American R&D. Without vibrant manufacturing and a research and development component, Colorado, a fairly small market, will have a harder time attracting more renewables, especially in a highly competitive market which is besieged by fossil fuels. The extension of the production tax credit for wind power has allowed the industry to survive but it has lost serious ground, continuing to lay people off. It is difficult to envision a long-term NEE without the green manufacturing component that it also envisioned.

Conclusion and lessons

The case study of Colorado's NEE provides three sets of important and comparable observations regarding sub-federal climate innovation policies in a US context: with respect to innovation, with respect to the interface of federal and sub-federal policies in a liberal capitalist economy, and with respect to the overall politics of EM in the US.

Innovations at the sub-federal level: Can strong climate innovation come at the sub-federal level (and by extension at the national level)? While it is evident that the federal government has played an indispensable role in fostering innovation in Colorado, there is clear evidence that state-level dynamics and initiatives were the key drivers of the institutional innovations that aimed at changing the state's economy. Were the participants aware of and affected by the broader context in which they operated? That is clearly the case. However, they sought to renegotiate Colorado's place in a particular and difficult direction. As one looks at the strategies of the various states with regard to energy and subsidies there is clear variation both in terms of what they prioritize and how (Story et al., 2012).

Moreover, this case demonstrates that it is important to look at climate innovation in a US sub-federal context not narrowly as a debate over technological change, but as a dynamic political process of institutional change that provides the context for future innovation. In this respect, we agree with the editors that policies are 'produced' by the institutional context and thus, policies and their impacts are *symptoms* of institutions or the 'rules of the game' that limit the range of possible outcomes.

Institutionally, the US is the *par excellence* liberal capitalist model. This key characteristic permeates all the way to the local level with noticeable variations, along a variety of issue areas from environment to labor (Lane and Wood, 2009; Almond, 2011). That the dominant liberal capitalist mode does not permeate evenly is due to the second important characteristic of the country – its federalism. Like other federalist countries such as Canada or India, federal-state relations have their own dynamics, formed over many generations. National policies are filtered down through that dynamic as sub-federal innovations are filtered up.

In either case, there is a third federal dynamic related to those above that drives innovation but also makes strong social and environmental reform at the state-level more difficult: that of regulatory competition. US states – and in fact agglomerations within states – are competing with each other and with locations, states and countries around the world. In some cases this is due to weak central governments, which want but cannot impose a national strategy. In other cases it is due to enabling central governments that either do not want or cannot impose a national strategy, not because they are conventionally weak, but because of the historical make-up of the country. This situation gives rise to a contradictory dynamic. On one hand, localities and sub-federal units are motivated to be leaders in innovation in order to enhance their competitiveness – whether to reverse decline, as is the case in Michigan, or move beyond booms and busts, as is the case in Colorado. However, in adopting innovations, these units, intentionally or unintentionally, produce certain trans-boundary outcomes. First, they may produce carbon leakage as fossil fuels are exported, for example, increased coal exports to China. If they are large enough, desirable enough, and able to, they can minimize this by making exit and entry difficult. For instance, if Colorado sought deep emissions cuts it could not only move to renewable energy infrastructure in state, but also, commensurably limit coal production and export. This is, of course, politically difficult. Or, as has been proposed, trade agreements and the World Trade Organization can include provisions that allow states and countries to exclude products that have benefitted from carbon leakage. Colorado has not shown an interest in this strategy, probably because it is too small an economy to have a meaningful global impact upon the architecture of international carbon accounting.

The second challenge is that of individual rationality and collective irrationality. As more and more US states and corporations compete for a new sector, in this case renewables, competition may foster abundance and cheaper production, but in the absence of a more coordinated strategy, the environmental benefits remain suspect.

Because Colorado adopted a 30 percent RES in 2010, even if it loses out in the renewables race, it will still be an environmental leader. However, if aggregate fossil fuel production increases 30 percent, renewables will be less consequential from a climatic point of view. Additionally, even if renewables take-off, there is a strong possibility that efficiencies will increases production and consumption of energy and raw materials – again making the 30 percent RES less consequential and thus, in isolation, a weak innovation.

Federal and sub-federal influences: Since its founding, the United States has seen a spirited internal debate, and a civil war, over the relationship between the federal government and state governments with regard to legal authority and resource allocation (Rabe, 2010). This debate has meant that explicit federal direction of economic policies has been eschewed in favor of more indirect and fragmented strategies with regard to national industrial policy and innovation. For much of the existence of the country, the federal government has played an active role with respect to infrastructure: transportation, communication and water reclamation. Its role in military organization and innovation picked up steam during the 20th century. After World War Two it became the engine of national industrial policy, joined by space exploration. Since the 1970s energy has joined in this repertoire. In addition, the federal government has enormous resources that it can use to channel private investment. But, at the end of the day, it has never had the centralized and directed strategies that coordinated capitalist countries have adopted.

States have long had their own economic policies. For instance, since the 1940s, the country has been divided between 'right to work' states which promoted anti-unionism as a comparative advantage and industrial states in which manufacturing unions were influential. Since the 1970s US manufacturing has been subject to global and national competition, with Southern states seeking to attract both global and national investment. The lack of strong federal economic policy has forced state governments in declining manufacturing states to assume many of the responsibilities associated with federal industrial policy. In this, they have to compete with extractive states, such as Colorado, and aggressive Southern states. One sector in which these states are competing is renewable energy manufacturing.

The federal government has not been a bystander in this process but has used existing organizations such as national laboratories, defense spending, and government agencies such as the DOE and EPA to effect the direction of renewables and energy development. As the most recent Energy Secretaries suggest, however, the federal government is supporting an 'all of the above' policy with wind and solar renewables being secondary to natural gas, biofuels and, most likely, nuclear energy. While climate policy seems to have risen again on Obama's agenda it is important to place it in this overall context towards energy. Those states that are interested in both renewables and climate policy, and are influenced by federal policies will reflect this contradictory energy/climate policy. Some of them may be more successful in using 'environmental federalism' to address climate change while others may simply engage in green energy manufacturing (Rabe, 2004). As this case shows, the direction of Colorado is very much affected by state-level politics and initiatives, even though the state is very much influenced by federal dynamics – both directly through NREL and indirectly, through changing energy markets.

Along these lines, the federal response to the financial crisis in 2009 saw allocations and subsidies for energy efficiency and the renewable energy industry in Colorado through the ARRA and the DOE's loan grant program constituting a strategic intervention on behalf of certain green priorities. Although this action provides the contours of an American green industrial policy, it is increasingly apparent that this may have been the apex of federal support for the near term. As ARRA money is exhausted and political support withers, hopes for further federal-level climate action have been dampened. It is important to note here that although DOE and ARRA policies have had an effect in Colorado, the lack of discernible evidence for coordination between federal and state-level institutions during this time period has required us to draw a distinction between the federal government and the NEE and the effect of their respective policies upon Colorado's energy landscape. The decision-making power for the allocation of federal loans, subsidies and stimulus has fallen outside of the institutional context of the NEE. This approach has enabled the federal government to directly target strategically important sectors of the economy during the crisis. Although debate over the efficacy of this approach persists, the allocation of federal resources towards a strategic economic end is undeniable. Likewise, its impact upon the states, although not always apparent, by design, has been a hallmark of this policy. It is in fact quite possible that this centralized federal approach may have actually harmed the institutional resilience of the NEE in the long term. For the state of Colorado, at least in the near term, we are more likely to see any further activity take the form of centralized federal policies rather than state-federal coordination with respect to policies designed to support climate innovation (Schrank and Whitford, 2009; Block and Keller, 2011; MacNeil and Paterson, 2012).

The prospects of EM American style: Innovations and sub-federal influences regarding energy, climate change or greening the economy can take many routes. It is possible to simply have an innovation strategy or an economic development strategy or an environmental strategy. Comprehensive attempts involving a decisive political leadership and a supportive alliance are rare in the US once one gets beyond specific activities where an 'iron triangle' may hold sway. The prospects of a national strategy of EM or green industrial policy for that matter, which received so much attention in the run up to the 2008 election, are not particularly promising now, despite Obama's assertions during his 2013 State of the Union address. In fact, the US is poised to become the largest producer of extractive energy in the world leading to increased investment in additional industries that benefit from fossil fuels.

Yet, Colorado offers an example of the possibility of EM American style, that is, one in which alliances are more fluid and institutional innovations seek to drive change at the level of corporations and other economic and social units. While the RES forces Investor Owned Utilities to change their mix it does not tell them how to do it. And while the Ritter Administration sought green manufacturing, neither it nor the federal government directly demands it. The likelihood of stronger EM in a liberal market economy is limited if not impossible for the same reasons that industrial policy is difficult. Such collective undertakings require commitment by all major parties to temper regulatory competition at the national-level, something that may be problematic in a liberal federalist context.

The fact that it is possible to move in the direction of EM at the state-level in a liberal market economy such as the US suggests that responses to climate change and the ecological challenge more broadly, are not mechanistically determined by the overall characteristics of a variety of capitalism. Even within liberal capitalism such as that of the US, it is possible to 'experiment'. At the same time, one should not underestimate the anti-environmental dynamics of liberal capitalism under conditions of crisis. The resurgence of extractive energy in the US and Colorado reiterates the difficulty of 'indirect' solutions to climate change, such as the RES. Long-term solutions require more profound innovations, that cap energy uses and emissions and diminish industrial and infrastructural practices that drive them. They also require innovations that limit emissions and contain carbon leakage. Barring attempts in the US to move towards a more integrated set of climate policies, with strong coordination between federal and state entities, it is difficult to foresee an effective response to climate change. State-level intervention is important, but insufficient in this regard. It is our conclusion that meaningful action on climate change will not occur within the current political context but rather will require a reformed liberal federalist capitalism, which has a greater institutional capacity to respond to the collective action challenge that climate innovation presents.

Notes

1 The chapter draws upon collaborative research conducted through a grant by the Colorado State University Energy Super Cluster with our colleagues Jon Fisk, Samantha McGraw, Linse Anderson and Professor Michele Betsill. We have relied upon seventeen interviews with key participants, primary documents, a wealth of secondary documents and publications and hours of research meetings that have resulted in a comprehensive and multifaceted study of Colorado's New Energy Economy from 2004–2012.
2 Suzanne Mettler defines the 'submerged state' as, 'a conglomeration of existing federal policies that incentivize and subsidize activities engaged in by private actors and individuals' (Mettler, 2011: 803).
3 In split estates the owner of the surface rights may not own the mineral rights. The issue is whether state law allows the owner of the mineral rights primacy over the owner of surface rights and what responsibilities they owe for surface damage. If mineral rights have primacy, in principle, a drilling rig can arrive one day demanding access to a specified location on the property.

Part II
The Corporate Context of Climate Innovation

In liberal capitalist economies, technological innovation is tightly tied to market demand. Part I demonstrated that market considerations dominate the US innovation system. Part II shows this to also be the case for corporations. As with government programs, financial considerations are the most significant determinants that a specific innovation will actually be produced and marketed. As noted in Chapter 1, the pull of market demand is measured by the anticipated contribution that the product or process innovation will contribute to the corporate bottom-line. As explained in Chapter 2, particularly in the US, managers accord shareholders primacy among all stakeholders as the ultimate determinants of their personal and their firm's financial success. However, to be effective much climate innovation involves the pursuit of long-term, radical advances in technology in which market demand must be, by definition, a small part of the reason why firms would invest in the required research. In the absence of a market case for investing in the research and marketing required, firms have to rely more on public sources of funding and regulatory support for radical innovation. The result is something of a paradox, but it logically follows that the more liberal the institutional basis of capitalism, the more the state must be called upon to perform those functions necessary to mitigate greenhouse gas (GHG) emissions that are not associated with delivering short-term profits and investment returns to shareholders.

In Chapter 5, David Levy and Sandra Rothenberg consider the manner in which the environmental problem of climate change came to the attention of corporations in the late-1980s and 1990s. Initially perceived as a contested scientific problem, it later came to be accepted as an economically strategic matter for corporations. The role of corporate scientists working in the US automobile industry illustrates how it did. This case study demonstrates that perceptions formed in the early stages of an issue such as climate change produce institutional responses that can have lasting impacts within corporations. More pertinently, they show that the scientists faced institutional pressures to view the scientific discourse through the 'frame' of market interest. These scientists are at the 'boundary' between science and corporate interests, and as such play a crucial role in 'translating' the science for senior corporate management. But the more removed these corporate scientists are from their scientific profession, and the greater their embedding in their companies' corporate institutional frames, the more likely corporate perspectives are to be shaped by a commercial logic. In other words, the preference for 'normal' market innovation is asserted over a rational and ethical response to the message of the science.

In Chapter 6 we focus on corporate justifications for action in respect of climate change today, now that the science of climate change has been accepted by much of the business community. We review corporate reporting by US firms under the Carbon Disclosure Project, and also report on interviews with senior office-holders from fourteen leading Australian corporations listed on the Dow Jones Sustainability Index. The corporations selected in both cases are from industry sectors that are highly GHG intensive, and therefore for which climate innovation is most crucial to climate change mitigation. We demonstrate that that the innovation process for liberal capitalist economies is unlikely to produce the necessary technological innovations without substantial government intervention. This is not because corporations desire regulation *per se*, but they would welcome clearly specified regulations, over the long-term, that allow them to embrace a business case for investment in climate innovation. In the absence of this, they perceive long-term uncertainty, and the market innovation that this encourages is very much 'business as usual', rather than the radical risk-taking that is required for climate change mitigation in advance of an environmental disaster.

Of course, the reports we analyzed and the interviews we conducted do not allow us to definitively divine the true intent behind many decisions that firms make with respect to climate change. Therefore, in Chapter 6 climate innovation is more pragmatically understood and measured as *technical change that reduces GHG emissions*. In other words, we take at face value firms' explanations of the purpose of the technical changes that they describe as reducing emissions. Even so, in most cases it is evident that firms are embarking on process or product innovations that reduce energy consumption which could be – and often are – primarily designed to reduce costs and increase profits. This is a small step in the right direction. However, in a liberal capitalist context, the government is the primary agent responsible for the 'rules of the game' that can provide the incentives for decarbonizing the economy sufficiently to avoid dangerous climate change.

In Chapter 7, Dimitris Stevis demonstrates why, even in a liberal economic context such as the US where they are in decline, unions have a potentially crucial role to play in driving institutional change for climate innovation. He shows that whether or not unions play a supportive role in this depends on many factors from the institutional context to internal culture of the organization. In the US, the last three decades have seen the decimation of union membership in some sectors of the economy. Unions have reacted to this in several ways but the most interesting response is that of the few that have chosen to collaborate with environmentalists in actively changing the political economy that they find so challenging. His is a novel perspective on labor relations in the most liberal capitalist economy and suggests that the pursuit of climate innovation may result in changing the balance of power between capital and labor in ways that economists have never considered.

The analysis in the chapters in this Part therefore expand on a key point made in Chapter 2: that liberal market economies cannot rely on the private sector to produce the necessary climate innovation without appropriately shaping the institutional context within which the market operates. The more corporations are market-focused, and the more this is accepted, the more the state must act in respect of what the market does not dictate, given that non-market environmental imperatives must take precedence if the predicted climate catastrophe is not to be realized.

5
The Role of Corporate Scientists and Institutional Context: Corporate Responses to Climate Change in the Automobile Industry[1]

David Levy and Sandra Rothenberg

Corporations are critical players in the worldwide effort to address greenhouse gas (GHG) emissions. They account for the vast majority of them, while also controlling the technological and organizational resources which, if applied appropriately, could play a major role in reducing GHG emissions. The science, technology and society (STS) literature examines the interface between science and policy and suggests that scientific knowledge and social structures of governance are co-produced, and that the boundaries between policy and science are inherently ambiguous and subject to continuous renegotiation (Kerkhoff and Lebel, 2006). The private sector, however, has generally been neglected in this debate. Although there has been some growing recognition of the role of private actors such as corporations in international environmental regimes (Clapp, 1998; Haufler, 1998), little attention has been paid to the role of the private sector at the science-policy interface. Yet, this role can be critical in the policy making process (Ehrlich, 2006).

What the STS literature tells us is that climate science does not simply land on the desks of policy makers and drive policy; rather, a complex social and political process mediates science and policy. In a similar way, the private sector is not a simple consumer of scientific findings and assessments. As one example, past research has highlighted the efforts of the fossil fuel industry to cast doubt on claims that GHGs are causing dangerous changes to the climate system (Gelbspan, 1997; Franz, 1998; Levy and Egan, 1998). Because corporations recognize that their economic interests are imperiled by potential measures to address climate change, challenging the science is a time-honored strategy for delaying or averting regulation (Jasanoff, 1990). These efforts are generally interpreted as strategic manipulation of scientific uncertainties and standards of proof. However, in this Chapter, we argue that corporate perspectives on climate science are not purely strategic, but are shaped by particular organizational structures, processes, and institutional pressures, and are internalized into the value and meaning structures of an organization. In turn, these perceptions of climate science influence a firm's organizational and strategic stance toward change. Similarly, perceptions of economic interests and climate science are mutually constitutive, as the perception of economic peril generates skepticism about the science, which in turn leads companies to defer investments in low-emission technologies.

We focus on the role of corporate scientists as boundary spanners during the early entry of the United States (US) automobile industry into the climate change science arena. This is because the automobile sector is responsible for substantial emissions, and is among the most prominent non-state actors in the emerging international regime to address climate change (Winter, 1998). How are the views of these scientists shaped by their personal position and networks and their perception of corporate interests? What is the influence of the company's history, structure, and culture, and its location in a particular national and industry institutional context? How do these scientists function as organizational gatekeepers and translators across organizational boundaries? Of most significance, perhaps, what impact do they have in shaping corporate strategies on climate change? To answer these questions, we first lay out how existing theory suggests that being on the boundary increases the likelihood that these scientists will introduce competing discourses into the firm, and thus be sources of institutional change. We then discuss how the frames that these actors use to interpret climate science can differ, with these differences due to both individual and contextual factors. Finally, we explore these theoretical arguments by examining the response of two US automobile companies, General Motors (GM) and Ford, to the climate change issue during its early emergence as an institutional field. The experiences of these two companies show us that these frames are influenced by their location on the 'boundary' between the corporate and scientific worlds. We conclude with a discussion of implications for policy and management, as well as opportunities for future research.

Institutional theory and corporate responses

Markets and organizations are embedded within institutional fields with important cultural, symbolic, and regulatory dimensions (DiMaggio and Powell, 1983). From an institutional perspective, then, corporate strategies regarding climate change are not purely objective economic calculations, but rather are based on understandings of climate science, expected regulation, and the market potential for mitigation technologies. These perspectives are, in turn, likely to be influenced by institutional actors, such as formal scientific assessments, competitors, industry associations, consumers, NGOs, regulatory agencies, the media, and scholarly journals. These actors constitute, in the language of institutional theory an *organizational field* which, over a period of time, establishes norms, policies, and standards of accepted behavior that shape a particular company's discourse and practices (Powell, 1991; Scott and Meyer, 1994).

Although institutional theory has generally been used to explain isomorphism of management practice among organizations, it can also be used to explain heterogeneity. Three primary theoretical arguments could account for heterogeneous organizational responses and perceptions. First, organizations often operate within multiple institutional fields, such as when they belong to different industry associations or operate in different regulatory contexts, creating divergent pressures (D'Aunno et al., 1991; Kempton and Craig, 1993; Alexander, 1996; Hoffman, 1997). The boundaries of such organizational fields are inherently unclear and may overlap or be nested in broader structures (Holm, 1995).

A second explanation for differences among companies is that an organizational field can sustain multiple competing *discourses*. Concerning the environment, one that many companies have adhered to is the notion that environmental regulations are inherently costly and antithetical to their economic interests. A growing group of companies, however, are embracing the discourse and practices of environmental management (Levy, 1997), termed 'eco-modernism' by Hajer (1995), who posits that incorporating environmental concerns into business strategy can reduce costs and build new markets. Within the automobile industry, the traditional discourse held sway until the mid-1990s, but thereafter an eco-modernist discourse contesting these conceptions provided an alternative 'vocabulary' with which firms engaged. Similarly, skeptical approaches to climate science provided discursive competition for those trying to use science to justify aggressive policy measures. These differing institutional logics are contested by different groups with different interests. Often, the resulting alternative discourses are introduced by *institutional entrepreneurs*. These are 'actors who create technical and cognitive norms, models, scripts and patterns of behaviour consistent with their identity and interests, and establish them as standard and legitimate to others' (Dejean et al., 2004: 743). These entrepreneurs have been identified as one source of emerging, competing discourse in institutional fields. In a review of institutional entrepreneurship, Leca et al. (2008) found that research pointed to three main conditions under which institutional entrepreneurs are likely to emerge: precipitating 'jolts' or crises; the presence of acute field-level problems that might precipitate such crises; and the degree of heterogeneity in the field.

These institutional entrepreneurs can also be *institutional interpreters*, as institutional discourses and practices do not pass undisturbed across organizational boundaries. Instead, they are 'translated' as they move across these boundaries (Boxenbaum, 2006; Boxenbaum and Battilana, 2005). Thus, these actors play a critical role in understanding heterogeneity as they construct the frames used in the process of institutional translation. Each company interprets institutional discourses through its own unique frame which is a product of its own institutional history and organizational culture (Lounsbury et al., 2003; Boxenbaum, 2006). The same holds for the individual institutional translators. For example, in her study of Danish business actors, Boxenbaum (2006) finds that different individuals translate institutional pressures in different ways, with translations varying according to individual personal and professional beliefs and interests. Delbridge and Edwards (2008) demonstrate that in the early stages of institutionalization, most work focuses on those who attempt to create change – that is, the institutional entrepreneur – and those who oppose it. However, it is just as important to understand the field, organization and individual factors that create the possibility for change to occur and form the context in which individuals interpret and initiate change (Delbridge and Edwards, 2008; Boxenbaum and Battilana, 2005; Tushman and Scanlan, 1981).

It is not that theories of institutional entrepreneurship ignore these aspects of context. Institutional entrepreneurs are, by definition, *strategic* actors. As seen in the work of Boxenbaum (2006), they translate institutional discourses altering their translations in a manner that would lead to greater resources, support and other factors essential for their own success. Similarly, Rothenberg (2007) finds that environmental managers alter the translation, or framing, of institutional pressures for improved environmental performance in order to increase the likelihood of response to these pressures. The institutional entrepreneur works within a context which may or may not provide the opportunities and consumers for the new discourses they are 'selling'.

Often described as the inherent paradox of institutional change, entrepreneurs also need to create change while their rationality is conditioned by the institutions they wish to change (Battilana et al., 2009; Beckert, 1999; Seo and Creed, 2002). This paradox is why boundary spanners are often sources of change in an organization. Boundary spanners are individuals or units who attend to and filter information about the organizational context and serve to link organizational structure to environmental elements, whether by buffering, moderating, or influencing the environment (Thompson, 1967; Aldrich and Herker, 1977; Leifer and Delbecq, 1978; Fennell and Alexander, 1987).

In answer to the paradox of institutional entrepreneurship, Seo and Creed (2002) argue that change within institutional environments is more likely to happen if an organization is exposed to multiple institutional fields and discourses within these fields, an occurrence that is more likely to happen at the boundaries of these fields. Greenwood and Suddaby (2006) argue that organizations which operate on the boundary of organizational fields are more exposed to field-level 'contradictions', resulting in lower levels of embeddedness and a greater propensity to act as institutional entrepreneurs. Therefore, it follows that individuals located at the boundary of their organization are the people most likely to be exposed to multiple fields and, thus, are more likely to be both institutional translators and entrepreneurs.

For corporations, the role of boundary spanners in the realm of environmental science can be filled by corporate scientists, whose job it is to collect and absorb scientific information from various sources (Furukawa and Goto, 2006). These scientists bridge the scientific and corporate worlds by working for a company while at the same time being members of their professional community. This membership is made manifest by publication in academic journals, participation in academic conferences, and membership in professional organizations, all of which encourage a discourse that supports the process of change (Furukawa and Goto, 2006; Greenwood et al., 2002). But for environmental issues where the science is highly contested, the science does not 'move naked from the lab or scientific journal into...boardrooms'.

Instead, what we see are 'elaborate representations of science' designed to show why one set of findings is better than the other Gieryn (1999: 4). For climate change, there has been great disagreement on the causes, extent and magnitude of the problem, including the specific rates of change and timelines and on global outcomes, as well as local and regional impacts (Rosenberg et al., 2010). Despite increased scientific agreement as to the seriousness of the issue, the extent of this conflict can be seen as recently as 2010, when email messages of climate scientists discussing ways to hinder the efforts of climate skeptics created a scandal that impacted international climate change talks (Satter, 2010). Adding fuel to this fire is that the climate change debate is value-laden, involving issues of equity, autonomy, and rights (Jiusto, 2010). The uncertain and contested nature of climate change leaves the door open for multiple interpretations and agency (Battilana et al., 2009). Because a company's perception of climate science can thus have a significant impact on their product, technology, and political strategies, corporate scientists play a critical role. For large multinational corporations their climate stance affects not only their own decisions but also the behavior of competitors, suppliers, policy makers, and other stakeholders. Given that society's response to climate change depends on strategic decisions taken by these companies, it is critical to understand how corporate perceptions of science develop and influence the industry.

The automobile industry's response to climate change

In the following analysis, we primarily focus on the experiences of two US automobile manufacturers, Ford and GM, in the late-1980s and 1990s. We chose these two companies because while they were similar in size (as compared to smaller Chrysler) and product ranges, yet they took different approaches to the climate science debate. We chose the timeframe because while there is a great body of literature on the responses of energy intensive businesses to climate change, most of it focuses on the later stages of institutional field development. Our focus on the earlier stages offers insight into how these firms acted while the science of climate change was developing and highly uncertain. Also, these early stages of translation can have a lasting institutional impact.

We utilized several methods for data acquisition. We conducted a total of thirty-three personal interviews over the course of several visits to the firms, primarily from 1998 to 2000. Interviewees included a cross-section of experts in the field of climate science and firm employees, including environmental staff, strategy, product development, marketing, and R&D. Most interviews were conducted in person, while a small number, particularly those that focused on the more historical data, were performed over the phone, using a pre-developed semi-structured interview format (see the Appendix to this chapter for the interview protocol). We gathered additional material through an extensive review of secondary source material. This included a Nexus/Lexus search from 1986–1999 using the terms 'climate change'; 'global warming'; 'greenhouse gas'; 'auto(mobile)'; and 'car' for the following news sources: *New York Times*; *Business Week*; *Financial Times*; *Wards Auto World*; *Auto News*; and *Wall Street Journal*.

We analyzed case material through coding and display of data. Interviews were coded using QSR Nudist (NVivo). We used two primary display formats, both of which are suggested by Miles and Huberman (1994) and Yin (1994). The first format is a temporal ordering of the data, in which specific events were placed in time lines created in order to gain a sense of historical development. The second format is a comparative matrix, in which segments were categorized and placed in a matrix in order to explore how cases, in this instance companies, differ from one another. We used our secondary source material to develop a general picture of the institutional and scientific background of the climate change issue.

Field emergence

The notion that human emission of GHGs might warm the earth's climate dates back to the work of Baron Jean Baptiste Fourier in 1827, Svante Arrhenius's first published estimates of the amount of warming caused by GHG-related radiative forcing in 1896, and establishment of an atmospheric carbon dioxide measurement station on Mauna Loa in Hawaii during 1957 as part of the International Geophysical Year. Scientific resources devoted to the issue grew rapidly during the 1960s and 1970s and policy makers began to turn significant attention to it during the 1980s. Figure 5.1 illustrates the number of articles mentioning climate change and the automobile industry in four of the leading outlets for business news – the New York Times, Financial Times, Business Week and the Wall Street Journal. As can be seen from this diagram, activity and interest of the auto industry in the climate debate only began to emerge in the 1980s and early 1990s. One of the critical events during this time in the US was the testimony of James Hansen before Congress in 1988 (discussed further below). This is similar to the notion of 'precipitating jolts', an enabling factor in the

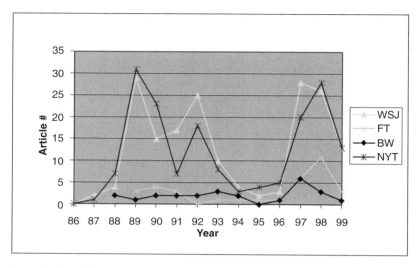

Figure 5.1 Articles mentioning climate change and the automobile industry, 1986–1999
Sources: New York Times, Financial Times, Business Week and the Wall Street Journal.

creation of change in institutional fields (Meyer et al., 1990; Greenwood et al., 2002).

Within a larger context of opposition to regulation of GHG emissions from industrial sectors related to fossil fuels, the American automobile industry has been among the most vocal opponents of mandatory emission controls, though it has not, of course, been alone in this opposition. The automobile industry's response has largely been channeled through industry associations, the most prominent being the Global Climate Coalition (GCC), an organization that represented about forty companies and industry associations, primarily major users of fossil fuels. GM, Ford, and Chrysler, along with the American Automobile Manufacturers Association (AAMA), were all members. A senior GCC staff member, discussing motivations for the creation of the GCC, expressed the view that industry had become involved late in the negotiations leading to the Montreal Protocol. As he expressed it, 'boy, if we didn't like the Montreal Protocol, we knew we really wouldn't like climate change! This is the mother of all issues!' A key strategy of the GCC and its member companies in its opposition to mandatory emission controls was to challenge the science of climate change, pointing to a spectrum of opinion rather than consensus among scientists, and highlighting the uncertainties.

The response of Ford and GM

The response of Ford and GM to climate change science closely mirrors the more general story just recounted. Despite the debate in the scientific arena on climate change and the participation of some internal scientists in this debate, corporate attention to the climate issue picked up speed only in the late eighties. Scientists at both companies were aware of the emergence of the scientific literature and at GM there was more widespread discussion of it prior to 1988. But climate change did not become a high-priority issue beyond GM's relatively autonomous research laboratories until 1988. At Ford, one manager described his shock at how quickly 'climate went from zero to sixty'. Thus, Ford's participation in the International Panel on Climate Change (IPCC) activities and more formal scientific scanning also started in the late eighties.

This delayed and 'surprised' response suggests that the automobile industry paid little attention to the development of scientific concern around greenhouse gases or to early interest in the policy community. The President's Science Advisory Committee had discussed greenhouse gases and climate as far back as 1965 and in the early 1970s two major scientific studies put climate firmly on the US policy agenda. The White House proposed a US climate program in 1974, leading to the National Climate Program Act of 1978, which authorized $50 million annually in research funding. The US Department of Energy initiated a CO_2 research and assessment program in 1977 and in 1979 the White House Office of Science and Technology Policy requested a study on climate from the National Research Council. The ensuing Charney report predicted global warming in the range 1.5 to 4.5 degrees Celsius over the next century. In 1983, the US EPA published a rather alarming report based on modeling by James Hansen from NASA. Yet, rather than the scientific evidence, management in the US automobile industry appears to have paid much closer attention to the considerable mass media coverage of Hansen's Congressional testimony, in which he stated he was '99 per cent certain' that recent warmer temperatures were attributable to GHG-induced climate change (Edwards and Lahsen, 1999). As one GM scientist recalled, 'we lived by the Wall Street Journal and the New York Times'.

Another reason cited by managers for the attention to climate in 1988 was the rapidity with which the ozone depletion issue had moved from scientific concern to the Montreal Protocol in 1987, mandating a 50 percent reduction in chlorofluorocarbon (CFC) production. Indeed, attention to CFCs in the mid-1980s might have diverted industry attention away from greenhouse gases. To the extent that corporate managers take their cues from the US institutional context and media (Levy and Kolk, 2002), with their rather parochial national focus, they would have been less likely to hear about major international conferences on climate. Although Detroit is closer to Toronto than to Washington DC, almost none of the managers interviewed recalled the June 1988 Toronto Conference on the Changing Atmosphere, which culminated in a call for a 20 percent cut in greenhouse gas emissions from 1987 levels by 2005. Even less known was the earlier series of workshops in Villach, Austria, held from 1980 to 1985.

While all three major US automobile companies, through their industry associations and independently, questioned mainstream climate change research and advocated a 'wait and see' attitude, they did so with differing intensity. GM was the only company of the big three to refrain from strong direct attacks on the science. Ford's Trotman and Chrysler's Eaton, on the other hand, were especially vociferous in the early 1990s, through speeches and editorials, in castigating concerns about climate change and emphasizing the high cost of precipitate action in the face of uncertainty. Interviews revealed that these views were not just those of top management but had permeated throughout various departments and management levels. One manager commented, 'we have followed the science as a company and we would like to see more science and less hot air! What we'd like to see is good science driving good policy'. As such, in the mid- to late 1990s, the automobile industry followed the GCC in focusing on climate model uncertainties. In a 1998 paper, Ford environmental scientists publicly attacked the science, stressing that the most significant oversight in current climate assessments appears to have been inadequate study of the role that the Sun may have played in climate change. They state (Petrauskas and Shiller, 1998: 6):

> Because of this, confirmation and quantification of the human capacity to influence climate beyond natural variability remains blurred. This fact alone does not completely eliminate all reason for concern, but it does loudly cry out for the scientific knowledge necessary to support far reaching global policy decisions... Real science needs to be verified first before such massive global changes in emissions ever could be justified in the future.

Change in tide

By the late 1990s, a shift in position was detectable in some sectors of the US industry. On June 8, 1997, the Business Roundtable sponsored full-page advertisements in the US press signed by 130 CEOs which argued against mandatory emissions limitations at the forthcoming Kyoto conference, citing scientific uncertainties and the high cost of action. In sharp contrast, an effort to coalesce an industry bloc supportive of emission reductions was led by the Pew Center on Global Climate Change in April 1998. Thirteen companies joined immediately, including Toyota. Pew then formed the Business Environmental Leadership Council, which signed on to a series of newspaper advertisements stating that they 'accept the views of most scientists that enough is known about the science and environmental impacts of climate change for us to take actions to address its consequences' (Cushman, 1998).

Although Toyota was the only car company to join the Pew Center, the US automobile companies also toned down their criticism of climate science as the December 1997 Kyoto international conference approached. According to the trade journal *Automotive Industries,* when the three US auto CEOs and UAW president Steve Yokich met with President Clinton in the Oval Office in early October of that year, 'they never questioned whether global warming was a scientifically proven concept' (Sorge and McElroy, 1997). Ford's Trotman recalled, 'we did not argue the science with the President. We didn't think that was a good use of his time or ours. It's generally agreed that the CO_2 in the air has increased in the last decades and that there's cause for concern, and that we should be doing something'.

A purely rationalist explanation for the shift in industry position would point to the emerging scientific consensus since the Second Assessment Report of the IPCC in 1995 and the strategic benefits for companies to 'board the train' once it was seen to be leaving the station. By 1997, the business press in the US and Europe was conveying the impression of consensus (Raeburn, 1997; Stipp, 1997; The Economist, 1997). From the perspective of a leading expert in this area, advances in basic science were fundamental to this change: 'the whole fingerprint argument has become much stronger since the SAR. You've got the empirical data of record warmth, and the arguments about satellite measurements and solar effects have been resolved in refereed scientific publications'. The growing body of scholarship in the field of science, technology, and policy should make us wary, however, of any simple linear connection between knowledge generated in the institutions of the scientific establishment, and societal responses. While the evolving science has surely played an important role in shifting corporate perspectives on the climate issue, the impact of this knowledge is mediated by the institutional environments in which companies are embedded.

Transformation of institutional pressures through organizational boundaries: Corporate scientists at GM and Ford

The corporate scientist as monitor and filter

Automobile companies construct perspectives on climate science based on multiple sources of information, both inside the company and in the external environment. IPCC-style assessments appeared to be somewhat discounted, however, as a source of information. As one environmental manager at Ford put it, 'IPCC reports had little effect, and caused few surprises. [Our internal scientists] already let us know what was coming down the pike'. An in-house scientist thought that her lack of attention to IPCC perhaps was due to its perceived political nature. She commented that 'the IPCC is the politics of science, not the science of science. I am inclined to stay out of the politics of it'. Instead, companies are more likely to obtain scientific information by interaction with outside academic experts. Both GM and Ford invited in a number of academic experts, though the list tended to be dominated by climate skeptics.

Companies might also obtain information through interactions with government agencies, particularly in the US DOE and EPA, and through participation in programs such as Climate Wise or the voluntary EPA GHG reporting system. However, again most of this climate science information in the automobile industry was formally filtered through internal scientists who acted as boundary spanners. Environmental scientists are usually trained formally in the sciences, such as atmospheric chemistry, and are actively involved with the external scientific community. GM, for example, employs a small group of environmental scientists who publish in peer reviewed journals, attend conferences, and participate in governmental panels. It is through their interaction with the scientific community that these internal scientists became the first employees to be aware of climate change as a potential concern for the firm. Thus, both Ford and GM used internal scientists, usually located in the R&D or environmental science department, to monitor the issue, filter and analyze the voluminous literature, and then translate the science through presentations and recommendations to management. It was the job of the internal scientists to relay the state of the science to others in the organization, and they therefore played an important role in shaping corporate perspectives on the issue.

With a large and independent research staff, GM appears to have been the first automotive company to follow climate science in a serious manner. The VP of Environmental Activities at GM heard about a 1971 scientific article concerning the role of particulates and greenhouse gases in the global climate system and he took an interest in their potential contribution to atmospheric cooling, the predominant climatic concern of the period. Ruth Reck, a scientist working in GM's research laboratories, was assigned to examine the question. It is interesting that awareness of the climate issue occurred through research involvement in other basic scientific issues, such as smog formation, tropospheric ozone, and CFCs, and particulates, in which GM and Ford labs were actively engaged. One VP of R&D at GM remembered that:

> Although most of the action had to do with tropospheric air pollution and emissions, there were several people who were real players in air mass movements and so on, so that there was a base of sophistication about atmospheric science systems. The first time it came onto my radar screen was in the 60s and 70s. I was certainly conscious of the worries that the world was about to freeze to death, so I thought I would follow it along. At GM it was around me as a developing issue, but that was more as a scientist than specifically as a manager.

Interaction with the scientific community, therefore, can be an important mechanism for early and continued awareness of the climate issue. This interaction takes a number of forms. The most commonly cited activity was the regular scanning of journals such as *Nature* and *Science*. Membership in scientific associations and associated activities also brought scientists in contact with the climate issue.

At Ford, prior to 1988, while Ford managers had held a number of discussions on the subject, they did not have anyone specifically assigned to tracking the climate issue. In 1988, however, after the Hansen testimony, Ford formally assigned an internal scientist to track climate science, and an engineer to monitor and participate in the negotiations over an international climate treaty and the IPCC process. This position was created when he advocated for his participation in major assessment processes, such as the UN negotiation sessions and the IPCC. As recalled by the VP of Environment, '[he] recommended that if we wanted to understand the human, political, and scientific dynamics of the issue, he really needed to be there'. Notably, he performed this external monitoring function not just for Ford but on behalf of the US auto industry and was funded and reported through the AAMA. As explained by Ford's VP of Environment, 'he was our window on the issues coming over the horizon'.

The particular role of these formal boundary spanners varies. Some of the information transfer is done on a more passive level, where the scientists serve as a resource rather than an active proponent of certain scientific concepts. When the issue becomes hot in the popular press, for example, scientists are often turned to for advice. Similarly, an internal scientist might be called upon to review material if an executive was going to testify to Congress or speak publicly about climate issues. Sometimes, however, internal scientists take a more active role in educating the organization regarding the state of climate science. In the case of ozone depletion, for example, Dupont finally reversed its stance only when its own scientists examined the theoretical and empirical evidence and concurred with external scientists about the gravity of the problem (Benedick, 1991; Rothenberg and Maxwell, 1995). This more active organizational stance taken by environmental scientists (both internal and external to the organization) represents their role as 'institutional entrepreneurs' driving change in their firms.

The clearest example of this more active approach was seen in General Motors. As noted earlier, Ruth Reck, a scientist working in GM's research laboratories, had been assigned to examine the climate issue. It is interesting to note that, at that time in GM, air quality was considered the more important and prestigious topic on which to be working. Nevertheless, Reck quickly became a world leading authority on particulates and on cloud formations, and was accepted into the closely-knit climate scientific community. She published in refereed scientific journals, and presented at numerous conferences and workshops. As chair of the first symposium on atmospheric chemistry in 1973, she actually turned down a submission from future Nobel laureate Sherwood Rowland on chlorofluorocarbons (CFCs), and later served as a reviewer for his landmark article in *Science*.

Reck, initially a climate skeptic, became an internal advocate for the issue by the mid-1970s, and also served as an important source of internal expertise, with regular access to top management. As remembered by Jimmy Johnston, GM's former VP of Government Relations, '[the environmental scientists] were very influential in putting the [climate] issues on the agenda. Ruth pushed what was really important, and was one of the more energetic people'. In an effort to alert the company to the climate issue and to find out what product divisions were already doing concerning GHG emissions reductions, she organized a large GM conference on the subject in 1985, which was attended by more than 700 company personnel. External climate scientists were invited to give presentations, notably excluding skeptics, whom she considered dishonest. Product managers were asked to speak about what they were already doing with respect to emissions and how this would be valuable in terms of reducing greenhouse gas emissions. Reck understood that she 'absolutely had to sell this issue', and used this and other company forums to that end.

The differing role of corporate scientists in the 'filtering' of climate science may help to explain differences in response between Ford and GM. At GM, where the corporate scientist was involved at an early stage with climate change research in collaboration with external scientists, the 'surprise' at the response to the Hansen testimony was much less pronounced. Similarly, while most US companies were taking a more 'wait and see' attitude to the science, GM, with an institutional entrepreneur who was an internal advocate of climate science, was the only company of the big three to refrain from strong direct attacks on the science. These differences are summarized in Table 5.1.

Distortion at the boundary

Despite their adherence to the scientific norms of objectivity and rationality, we found that with the exception of Reck, the internal scientists tended toward the skeptical end of the spectrum of legitimate opinion among respected climate scientists (Morgan and Keith, 1995). They all interpreted scientific uncertainties in a conservative manner, viewing them as a rationale for further research rather than seeing the potential for climate shocks from positive feedback or threshold effects. They pointed to the long time frame of atmospheric accumulation of GHGs as signaling a comfortable margin of time for reducing uncertainty rather than an urgent reason for early precautionary action. Thus, the predominant voice within the automobile companies was one of skepticism that climate change was a major concern requiring significant private investment or government regulation.

To some degree, these perspectives can be attributed to strategic posturing. Yet the skepticism toward climate science that we encountered across many interviewees at various levels cannot all be understood in

Table 5.1 Ford versus GM's interface with climate science, 1970s to late 1990s

	Ford	GM
Date corporate scientist specifically assigned to track climate issue	1988	Early 1970s
Company response to Hanson testimony	'Shocked' and corporate scientist began to participate in IPCC, which was called 'the politics of science'.	Less shocked, but still mobilized wider corporate attention.
Engagement of corporate scientist with scientific community	Participation in the International Panel on Climate Change, partly representing the AAMA.	Highly respected researcher accepted by scientific community. Purposely did not represent GM in scientific activities.
Role of corporate scientist	Was 'our window on the issues coming over the horizon'.	Was the 'clearinghouse for information' and understood that she 'absolutely had to sell this issue'.
Invited speakers	Invited in a number of academic experts, though the list tended to be dominated by climate skeptics.	Internal conference on the subject in 1985 where external climate scientists were invited to give presentations, *notably excluding skeptics*. Skeptics were brought in at other times.
Company response to science	Public attacks on climate science. Trotman was vociferous in early 90s in castigating science. Paper presented by corporate scientist in 1998.	GM was the only company of the big three to refrain from strong direct attacks on the science.

such terms. While managers in departments responsible for public and government relations might have been comfortable 'spinning' the science in a particular way, there also appears to have been a process of internalization of these perspectives. We came away from our interviews convinced that most managers sincerely believed in the skeptical position. While Johnston admitted that the adversarial political system in the US required some strategic exaggeration, and as one climate scientist in GM recalled '(Johnston) had to assume this position because he was the chief lobbyist', it was clear that he was sincere in his skepticism about the science and the role of government regulation. After retiring from GM, Johnston joined the American Enterprise Institute where he wrote a book about his experiences (Johnston, 1997). In order to understand how this distortion occurred, we need to look at the context in which the scientists worked.

The scientist in context

The process by which these conservative viewpoints are institutionalized is complex and not directly evident from interview responses. One person suggested that there might be some element of self-selection in terms of who is willing to be a corporate scientist. This is in line with Noble and Jones (2006), who found that boundary spanners are selected based on having the skills and abilities of an effective boundary spanner, rather than volunteer for the job based on their interest in the task. Another person who worked closely with GM on these issues commented that it might have to do with where they get their information: 'if they are reading GCC literature and the Wall Street Journal, then they get a particular impression of the issue'. Lastly, the managers and scientists worked within an organization that felt threatened by the prospect of regulatory action to address climate change. One executive discussed the pressure to adopt a bottom-line perspective. She recalled that there was a need for credibility with the 'line guys'.

There appears to be a subtle process of negotiation of identity between perceptions of corporate or departmental interest and an individual's own viewpoint. Psychologists have long observed that people are averse to 'cognitive dissonance', holding conflicting ideas simultaneously, or conflict between action (such as work routines) and ideas (Festinger, 1957). So people who work for a car or oil company can reduce their internal conflict, or dissonance, by embracing climate skepticism. As one ex-R&D manager expressed, 'there is social pressure. For the [internal scientist], they are around people who don't pay attention to the climate issue and don't want to hear it.... People on the operational side are more conservative'.

The 'location' of these scientists on the boundary between the corporate world and the scientific world seemed to be important in determining the degree to which the scientists were influenced by these pressures to adhere to a more skeptical perspective on the science. Reflecting the tensions of their location, corporate scientists strive to adhere to the norms of objectivity, rationality, and free investigation while being embedded in the business culture of bottom-line accountability and hierarchical subordination. This bridging of two cultures necessitates a subtle process of negotiation of identity for these scientists, who are not quite at home in either setting. The degree to which they are embedded in organizations on each side of the boundary has an impact on this process of negotiation, so that the extent to which a boundary spanner interacts with people and organizations on each side of the boundary can influence the level of identification with these parties (Finet, 1993; Richter et al., 2006).

Thus, for Reck, who was clearly immersed in the scientific discourse and the professional activities of the scientific field, the importance of identifying with the scientific community was clear. It was very important to her that she maintained her identity as a scientist and behaved in ways that would signal this to the scientific community. She reflected, 'I was an independent scientist. I have refused to be bought my whole life.... It is a strange hybrid culture being a corporate scientist'. This stands in stark contrast to Ford scientists who were relatively insulated from university or government-based scientists and who presented a scientific paper at the World Automotive Conference in 1998 that directly questioned climate science (Petrauskas and Shiller, 1998).

This influence of context becomes more complex when you examine it in more detail. While companies attempt to speak with a single authoritative voice in public or to regulators, there were frequently significant internal tensions over controversial issues. These tensions had bedeviled the development of GM's electric vehicle during the early 1990s (Shnayerson, 1996). The influence of material interests on perceptions of climate science was particularly stark in comparing differences across departments and functional areas. For example, managers responsible for advanced automotive technologies, including hybrid and electric powered vehicles, tended to see climate change as a real problem which presented an opportunity for product innovation. The spirit of the research labs tended to be 'we will show top management we can do it – we can change things'.

While the R&D people had a vested interest in developing solutions to problems and tended to view these solutions as technologically feasible, others in the organization were likely to take a more conservative approach. Managers in government relations and regulatory affairs departments, in particular, have traditionally seen their jobs as opposing governmental regulation and mandates. These managers were frequently concerned that the company might encourage more stringent regulation by demonstrating technological capabilities for reducing emissions, even if these technologies might be costly and unappealing to consumers. Managers responsible for traditional product divisions and strategy were particularly concerned about the high cost of low-emission technologies with little direct value to consumers. These managers clearly understood that both Ford and GM enjoyed a competitive advantage in the large vehicle and truck segments of the market, so they were particularly vulnerable to emission regulations that would raise fuel prices and encourage demand for smaller vehicles.

Again, given these competing interests the organizational location of the internal scientist, in addition to their location on the boundary, becomes very important. The greater the level of accountability of the scientific staff to other functions such as product management, marketing, and government affairs, the stronger the institutional pressures to conform to the climate skeptic position (DiMaggio and Powell, 1983). By contrast, Ruth Reck, the strongest proponent of climate change, operated in the GM labs for the most part as an independent researcher, evaluated as an academic rather than as a business manager, with promotion dependent on external publications.

The role of the internal scientist as institutional entrepreneur is also dependent on access to the power and resources needed to effect change (Kochan, 1975). Reck's unique access to information, for example, put her in a strong position in this regard. She recalled, 'I was the only one working on climate. Everyone in the corporation had to come to me, as I was the clearinghouse for information. This was a powerful position'. But this status could be a double-edged sword. Scientists such as her were also seen as remote from the core profit generating activities of the company, and their location in R&D labs or headquarters staff tended to isolate them somewhat from managers with line responsibility for product design and development. This was particularly true for the 'research' arm of the R&D departments, whose research may or may not relate directly to near-term product development. As explained by one R&D manager, 'most of the time the R was separate from the D. 97 percent of the R was in [the] laboratory. D was sort of a molecular film spread out over the company'. This was particularly true in GM given its highly decentralized structure, with the relative autonomy of GM's R&D and basic science leading to an overall perception that the scientists were not contributing to the needs of the firm. Therefore, while decentralization of R&D enabled corporate scientists to pursue their interests in climate and maintain a degree of autonomy, it may have also reduced the ability of these scientists to influence corporate policy or product strategy.

In recognition of this problem, there have been efforts to further integrate environmental scientists with the rest of the organization at both companies. As the climate issue gained in prominence, for example, GM realized that it needed a direct bridge between their climate scientist and their policymaking processes. Reck was directed to liaise closely with the executive VP for government and public affairs. This was a rather odd combination, given that the scientist had been the internal champion of the climate issue, and the VP's job was to convince the public and the government that mandatory emissions reductions were unnecessary and economically disastrous. Therefore, while this move gave Reck a voice near the top of the company, paradoxically it constrained her ability to promote the issue internally.

In the early 1990s, there was also an effort to increase the market relevance and accountability of research, and research projects were required to gain the sponsorship of a product division. In addition, both GM and Ford substantially eliminated basic scientific research during this period, reasoning that they should not waste their money generating non-proprietary knowledge to be disseminated in journals. The argument was that universities and government research laboratories were better positioned to conduct this research and companies could draw from this body of expertise when necessary. This integration shifted the balance for corporate scientists from an academic culture to a more traditional corporate setting. By the mid-1990s, neither Ford nor GM had internal scientists who were major players in the climate science community. This shift may also explain the relative skepticism of corporate scientists by the mid-1990s.

Not only must internal scientists negotiate inside a firm with multiple coalitions, they also need to balance their business role with the one they play in the scientific arena. The corporate scientists interviewed were particularly emphatic about their objectivity and independence, relating stories to demonstrate their refusal to be curtailed by narrow corporate interests. Ruth Reck was on an EPA advisory committee, and in her words, 'GM desperately wanted to remove me from it.

They thought I was not toeing the GM line. But...I was never on anything representing General Motors'. Instead, she was on the committee as an independent scientist. Although not threatened with her job because of her independence, Reck knew there was dissatisfaction with this role. With their loyalty to the corporation in some doubt, corporate scientists needed to negotiate the border between these two cultures with some careful diplomacy. Reck recalled that 'you had to speak strictly in terms of facts. Lots of people got into trouble for saying controversial things. I lived by the rule that anything you say might appear on the front page of the New York Times. Anything I said could always be backed by a reference'.

If corporate scientists were not completely at home in the business world, it was also the case that participation in EPA and IPCC panels was viewed within the corporate culture not just with suspicion but as a waste of valuable corporate resources. A number of scientists mentioned that external activity was viewed as unproductive and that their corporate departments and operating divisions were reluctant to bear such a 'tax'. In this atmosphere, it is not surprising that the IPCC has had difficulty recruiting authors from industry, despite IPCC chair Robert Watson's redoubled efforts to do so (Watson, 1999).

Institutional histories and leadership

It is important to recognize that the skeptical discourse concerning climate change is rearticulated within each company in the context of its particular institutional history, market position, and corporate culture. For example, GM had invested an estimated $500 million in the development of its electric vehicle during the 1990s, but less than 1,000 had been sold. Although a few GM managers thought that the company had gained valuable expertise in electric drive trains, the experience was widely perceived as a commercial mistake. Similarly, GM managers felt that they had rushed too quickly to downsize their vehicles, particularly luxury vehicles like the Cadillac. The industry perception was that Ford was making more money in the late 1990s because it had maintained its full-size vehicles and expanded production of trucks and SUVs faster than GM. In this context, the perception of climate regulation as a major threat became the dominant perspective in GM. Ford managers also recounted a piece of company history to explain their collective wariness of low-emission vehicles. The firm had invested an estimated $500 million in sodium-sulfur batteries, only to abandon the project because of safety concerns and because nickel metal hydride looked more promising.

The responses to climate change also need to be placed in the context of the 'siege' mentality prevalent in the US automobile industry overall. A senior executive at GM noted: 'there is a broad agenda of efficiency, and against large vehicles as poor choices. Climate is only the most recent driver of fuel efficiency. Before that there was oil dependency, the energy crisis, urban sprawl, and smog'. Johnston put it in more political terms: 'there are people who have cast the automobile as a villain. It is a puritanical view, that we are having too much fun, that we have too much mobility and freedom, that suburban sprawl is bad. They think we should all live in beehives. So when scientists say that CO_2 is a greenhouse gas, they jump on board'. This mindset tended to make managers see climate change as part of a larger ideological and political attack on the industry, justifying their skepticism. At the same time, managers saw the continuing pressure and momentum toward fuel efficiency and carbon regulation, and felt compelled to accommodate them to some degree.

National context

The debate over the strategic value of challenging climate science also needs to be understood in the context of the particular nature of the science-policy interface in the US. The congressional hearings on climate exemplify the adversarial, legalistic courtroom style through which the scientific basis for regulation is developed and contested. This contrasts sharply with the more integrated, consensual approach found in Europe (Mikler, 2009; Jasanoff, 1991; Edwards and Lahsen, 1999; Kruck et al., 1999). The institutional governance structures in the US causes companies engaged in contested policy arenas to make their case in a vociferous, public manner. As one auto executive put it, 'the Hill works by compromise, so you need to go to the extreme. The more strident one side gets, the more the other side must. It ends up completely polarized'. GM's Johnston made the case that the automobile industry ended up with Corporate Average Fuel Economy (CAFE) in the first place because of a misplaced strategy of conciliation: 'the Neville Chamberlain approach doesn't work. We offered voluntary fuel economy goals, with the usual rationalization that the train is leaving the station and we have to get on board. But the government turned around and made it mandatory. If we offer to do X, they will demand two X'.

The geographic structure of US-based auto companies reflects a national basis to their organizational fields, which might also be a factor in the conservative response of the US companies. Although these companies have been multinational for many years, the orientation of top management is primarily domestic. In numerous interviews, corporate managers, many with worldwide responsibilities, spoke about the difficulty of reducing emissions with gasoline at $1 a gallon, consumers who care little for fuel environment and are hungry for large SUVs, and a Senate unlikely to ratify the Kyoto Protocol. These views were reinforced through membership in industry associations dominated by US-based companies. This situation only began to change for Ford and GM during the mid-1990s, as Ford implemented its Ford 2000 project which pushed toward the rationalization and integration of production and management worldwide, and GM began to move in a similar direction. By 1998, top management in both companies included a number of people with significant overseas experience.

Conclusion

How companies perceive the risks of environmental threats such as climate change can have a significant impact not only on their own behavior, but also the overall dynamics in the larger institutional field. Perceptions formed in the early stages of institutional change are particularly important since they can have lasting impacts. In this chapter, we aimed to better understand the role of the corporate scientist interpreting and filtering the discourse on climate science, thus influencing corporate perceptions of this science. Moreover, we wanted to understand the factors that influence this process of translation.

As expected, we found that the context in which these scientists work influences the frames they use to interpret this discourse. In particular, we found that the placement of these scientists on the organizational boundary is critical. Corporate scientists need to balance their business and science roles and the corporate scientists interviewed were particularly emphatic about their objectivity and independence, relating stories to demonstrate their refusal to be curtailed by narrow corporate interests. However, the more independent organizational location and external professional orientation of GM scientists at the time climate change was first being acknowledged enabled them to engage earlier with the climate issue and become internal advocates for change. With this identification, the norms of objectivity, rationality, and free investigation played a larger role as the scientists attempted to translate the 'truth' for their organizations despite facing institutional pressures to view the scientific discourse through the frames of market interest. As a result, GM's product, market and political strategies were more appropriate to the climate issue then emerging.

We also found that despite the tendency to institutionalize conservative and skeptical perspectives on climate change, at the very top levels in both companies there was a genuine concern to 'know the truth'. As one put it, 'the trick from a management standpoint is how to get information through the layers of the organization and be able to make a judgment. We want to know what's really going on, not just what we want to hear'. Managers acknowledged that if the more pessimistic forecasts were borne out, the Kyoto commitments would need to be substantially strengthened, with drastic implications for the industry. Another ex-VP commented on how top management prefers certainty, even if the news is unwelcome. Recalling the story of DuPont and CFCs, he stated:

> the head scientist came back and said 'guys I am convinced it is real.' Then DuPont could move. In a sense, if the scientists were able to say 'I saw yesterday's data and it's certain' the industry would breathe a sigh of relief...but as it stands, we are uncertain about the science and what the politicians are doing.

In an ironic twist, top management expressed a sincere desire to understand the true scope of the climate problem in order to make strategic plans, yet the automobile companies and the scientists in them, were to varying degrees constrained by institutional perspectives that reflect the perceived threat to their interests. Corporate scientists do not deliberately distort the scientific literature, but our research does suggest that, through their role as filters, monitors, and advisers, embedded in their companies they are constrained to not offer a full spectrum of opinions.

Given the focus of this study on two American automobile companies, the question remains if these findings are generalizable to a broader range of companies, particularly those companies that operating in different national contexts. The study is also limited in that it only looks at the early stages of institutional change. In the later stages of institutional change, the discourse within the scientific community changed as consensus grew among scientific community. It is not clear what role corporate scientists would play in these later stages. Another issue for consideration is the translation process for companies that do not have dedicated environmental scientists, a likely scenario as companies increasingly outsource their basic research (Quinn, 2000). It is possible that these companies would turn to industry associations or to external university-based scientists, who might provide very different perspectives. The boundary spanning role for environmental issues has historically been given to the Environmental Health and Safety department, though increasingly to sustainability teams. These are frequently established as a response to a demand for socially legitimated roles. Under this situation, as found by Rao and Sivakumar (1999) in the establishment of investor relations departments, it may be that the boundary spanners act more as buffers or simply signal to the external environment a corporate engagement with environmental issues. These managers would not be professional scientists and are unlikely to serve as internal advocates for change. They follow a different set of 'scripted roles' that are enacted by themselves and their spectators, such as the media and government (Lamertz and Heugens, 2009). In the adversarial US policy making process, these roles call for exaggeration and contestation.

Overall though, the importance of corporate scientists in interpreting institutional discourses around environmental issues is clear, particularly in the early stages of institutional change. Some suggest that the environmental issues of the future will be similar to climate change in that they will be global, value laden, and fraught with uncertain science. Thus, there is much to learn from this phase in the change process. We found that the less embedded the scientist is within the external scientific profession, the greater likelihood that the interpretation process is shaped by their organization's institutional frames. The result is that corporate perspectives on key scientific issues are largely shaped by operational and commercial logic. The risk for companies is that these market logics are conceived and institutionalized in the context of historical experience rather than future pressures arising from issues such as climate change.

Appendix: Interview Protocol

1. Tell us about yourself and history with your company.
2. Can you explain how/why your strategy regarding climate change has evolved?
    a. How has this strategy affected your research and development priorities? Which technologies are you most optimistic about and why? How important is it to develop these two technologies in-house or can they be acquired from outside? Which are the most promising?

b. How did your company's stance with regard to government policy change over time?
      i. Probes: tolerance of a mandatory carbon trading system, higher CAFE standards in the US, standards in the EU?
      ii. How was this influenced in your firm's recognition of climate change as a problem?
3. Where does the company (you) get information on climate science?
4. Does the perception of the science vary within the organization? How? Why?
   a. Has there been a convergence of climate change strategies between North American auto companies and European ones? If so, what is behind this?
5. Has your company shifted its climate related institutional affiliations over time?
   a. Joined any climate initiatives?
   b. Developed relationships with external environmental and scientific organizations?
   c. Become more involved in the IPCC process?

Note

1 This chapter is adapted from a previously published article, Rothenberg, S. and Levy, D.L. (2012), 'Corporate Perceptions of Climate Science: The Role of Corporate Environmental Scientists', *Business and Society*, 51(1), 31–61.

6
Corporate Investment in Climate Innovation

Neil E. Harrison and John Mikler

As the majority of technological innovation is generated in the private sector, albeit supported by government funding, it is broadly accepted that climate change mitigation depends on corporate decision-making (Newell and Paterson, 2010). What incentives would persuade firms to generate the necessary incremental and radical technologies? The answer is usually along the lines of 'getting the prices right' by taxing greenhouse gas (GHG) emissions; introducing market-based systems to trade GHG emission rights; and subsidizing the creation and diffusion of new low-emitting technologies. Such policy prescriptions spring from a particularly liberal conception of capitalism in which markets solve economic coordination problems, and this is evident in both Australia and the US. In Australia, market mechanisms were the main policy option for GHG emission reductions debated for over a decade before a price on carbon, or a 'carbon tax', was ultimately introduced in 2012 (Crowley, 2013; Christoff, 2013). In the US, a strong undercurrent of neoliberal ideology similarly frames all social, environmental, or economic challenges in terms amenable to market solutions (see also McGee's analysis in Chapter 8).

In this chapter we consider the need to focus on the national political economy of climate innovation, focusing on the US and Australian contexts. We present an analysis of the responses of large US-based corporations across eighteen industries to questionnaires from the 2012 Carbon Disclosure Project (CDP), and the findings from interviews conducted with key office-holders from Australian corporations listed on the Dow Jones Sustainability Index (DJSI) immediately after the introduction of Australia's carbon tax. Both samples are taken from industrial sectors that are the most GHG intensive, and therefore where climate innovation is crucial.

We draw two principal conclusions from our research. First, we find broad support for our hypothesis in Chapter 2 that the market innovation process is unlikely to produce the necessary technological innovations without substantial government intervention. Major US firms accept the science of climate change and anticipate more regulation, but most currently plan only incremental increases in energy efficiency primarily by modifying their production and operational processes. Despite the introduction of a tax on GHG emissions in 2012, the Australian interviews suggest that this in isolation is unlikely to be enough to drive long-term planning for innovative products that would significantly reduce GHG emissions. Secondly, both US and Australian firms are less concerned about the type of regulation than its stability. The long-term planning horizon necessary for more radical climate innovation, means that the regulatory uncertainty produced by politically motivated short-term policy changes effectively stifles it.

Thus, our analysis challenges the conventional wisdom that industry is naturally reluctant to accede to government intervention to mitigate climate change (for example, see Hamilton, 2007; Pearse, 2012). In fact, we show that both the CDP reports of large, US corporations and the interviews with major Australian GHG emitters demonstrate the need, and corporate support, for *more extensive* and *stronger* government intervention, provided that this is strategic, long-term and appropriate to the challenges that firms face.

Our chapter proceeds as follows. First, we consider the ineffectiveness of firms' decisions on mitigating their GHG emissions. Analysis of both global and US emissions suggests that firms' efforts are insufficient to achieve global targets for avoiding dangerous climate change. We then present our analysis of reports by forty-seven firms to the CDP in 2012. These firms used nearly 2.8 billion megawatt hours of energy and emitted 837 million metric tonnes equivalent (MTe) of CO_2 in earning more than \$1.7 trillion of revenue (approximately 1.87 percent, 2.65 percent and 2.1 percent, respectively, of global totals). Our analysis of their reporting demonstrates that they are essentially waiting for a reduction in regulatory uncertainty, and in the absence of this almost none are investing for the long term. In the third section we report on interviews with fourteen leading corporations across Australia's most carbon-intensive industry sectors. As in the US, the Australian firms' interviewees indicate that their companies would welcome regulation if it reduces long-term uncertainty. Finally, we conclude on what these findings imply for climate innovation in a liberal economic context: more creative government intervention that is stable over a longer period of time.

The ineffectiveness of corporations' climate innovations

Liberal capitalism is based on the presumption that collective objectives can be met through the individual pursuit of self-interest by rational economic actors such as corporations. Similarly, market failures are seen as being correctable by 'fiddling' with market incentives they face. For example, national GHG emission reduction targets should be met by firms individually choosing to reduce their energy consumption when fossil fuel energy prices are high enough. At the same time firms would invest in redesigning their products and services to use less energy as consumers demand greater energy efficiency (for example, see OECD, 2009; and more critically Koch, 2012).

The CDP takes a different, though still market-focused, tack. Rather than offering economic incentives in the form of input cost increases or consumer demand, it is designed to use the heft of major institutional investors to persuade firms to invest in rapidly reducing their GHG emissions. The CDP, like other efforts to institutionalize carbon disclosure, is 'a political project because it entails a change in the structures of corporate governance' to inject consideration of climate change into corporate strategy and management processes (Kolk et al., 2008). It is a form of 'civil regulation'. However, in order to draw in investors the CDP narrative is centered around reducing investor risk from undisclosed climate change threats at potential investee companies. The chain of causal logic employed is that 'carbon reports need to be relevant and valuable to investors by conveying information that relates to the financial impact of climate risks and carbon controls on the valuation of corporate assets'. Thus, the CDP claims the data show that the leading corporations for climate change disclosure and performance provide greater returns to their investors, so by organizing and strategizing for climate efficiency corporations improve their return on capital employed. Climate efficiency equates to capital efficiency, and the hope is that the leverage of institutional investors who realize this will persuade more firms to monitor their GHG emissions, include climate change as a substantial part of their corporate strategy, organize their internal processes to reduce emissions, and increase investment in climate innovation.

By 2012, institutional investors controlling $78 trillion had signed onto the CDP and received access to all its reports (CDP, 2012). Yet, there is little evidence that either investors or companies have significantly changed their practices. The managers and shareholders who dominate corporate governance are prone to oppose 'intrusions into their autonomy', and complain of 'the costs of reporting' and the potential loss of market share (Kolk et al., 2008). Despite optimistic assessments that 'financial markets are starting to reward companies that are moving ahead on climate change' (Cogan, 2006) there is little evidence that the CDP reports provide data actionable by investors (Kolk et al., 2008).

For anyone concerned about the threat of a changing climate, analysis of the publicly reported strategies and processes of the world's 500 largest firms, one-third of which are headquartered in the US (Wilks, 2013), suggests that progress to avoid dangerous climate change has been insufficient. Even though nearly all the CDP reporters accept the science and recognize the potential risks and opportunities of climate change, most are not doing enough to forestall it. 96 percent of them have established board oversight, are reporting on their efforts, and on average are reducing their total GHG emissions by about 1 percent per annum. But the reductions agreed at the Seventeenth Conference of the Parties to the UNFCCC in Durban require a 4 percent annual reduction between 2020 and 2050. Analyses such as that of PWC Advisory Services (2012) similarly find that the global rate of annual de-carbonization required to avoid a 'dangerous' average 2°C rise in global temperature is 3.7 percent from 2000 to 2050 (PWC Advisory Services, 2012). Only 20 percent of firms have even set emission reduction targets beyond 2020, and whatever the reductions they have achieved, only 40 percent of reporting firms admit that this was strategically intentional. About half the CDP reporting firms see opportunities in adapting to climate change, but only a fifth have set aside an R&D budget for climate innovation to mitigate it (CDP, 2012).

US performance has notably lagged the global target. Between 2010 and 2011, the US reduced its energy-related absolute emissions by 1.9 percent in a slow-growth economy, and its carbon intensity declined by 3.5 percent. However, because its carbon intensity has only fallen an average of 2.1 percent per year since 2000, it now has to de-carbonize at a rate of 5.2 percent annually through 2050 to meet its presumed share of the global target. A handful of corporations, the thirty-four members of the Carbon Performance Leadership Index (CPLI), have recognized the long-term strategic importance of climate change, the risks it brings and the opportunities it offers. But US firms in particular are laggards in both reporting and performance. Only eight US firms meet the stringent criteria and are included in the CPLI, not a single US firm scored in the top ten for reporting and performance, and only three US firms scored an 'A' grade within their sectors for disclosure and performance (CDP, 2012).

The Obama Administration recently committed to a 17 percent reduction in absolute carbon emissions from 2005 levels by 2020 (Executive Office of the President, 2013). Because of Congressional opposition, this is not a legally binding commitment and it is substantially less than suggested by the IPCC. An inability to achieve even this modest target would remove any possibility that the US could continue to influence international negotiations in its favor, let alone persuade China and other newly emerging market economies to participate in reducing GHG emissions. Though some optimistic analyses exist (such as Burtraw and Woerman, 2012), if the GHG emissions of the 100 largest US companies between 2007 and 2009 are any indication, the US has little chance of meeting its target. Over that period, their emissions actually increased by 0.36 percent annually (CDP, 2010).

Climate innovation in the US: CDP reporting

In Chapter 1, we defined climate innovation as technological innovation with the primary purpose of mitigating climate change. In our review of the CDP reports we are usually unable to divine the true intent behind many decisions that firms make with respect to climate change. Therefore, in what follows climate innovation is more pragmatically understood and measured as *technological change that reduces GHG emissions*. To a large extent, we take at face value firms' explanations of the purpose of the technical changes that they describe as reducing emissions, but even so, as will become clear, in most cases it is evident that firms are embarking on process or product innovations that reduce energy consumption which could be – and often is – primarily designed to reduce costs and increase profits.

The CDP distributes questionnaires to more than 1,500 leading companies worldwide. The main questionnaire has fifteen sections, nine of which are concerned with the methodologies and results of emissions measurement, that we largely ignored, and six sections of substantive questions on the impact of climate change on governance practices, strategies, and their subjective perceptions of the risks and opportunities that climate change presents. We refer to the published responses to the questionnaire as 'CDP reports'. As outlined in more detail in Appendix 1, we selected 47 CDP reports across eighteen industrial sectors. We chose the sectors either because they have historically been large GHG emitters or because they are likely to generate climate innovation. In most cases we selected all the reports publicly available for each sector. The firms included have also self-selected as leading reporting entities and, therefore, those most likely to be advanced in the integration of climate change into their governance, strategies, and management processes.

Corporate governance

All the firms in our sample accepted the climate change science. Like American Electric Power, many stated that human activity 'has contributed to global warming' and 'are discovering that many of the lines [they] had drawn separating financial from nonfinancial strategies, activities and reporting are no longer relevant and, in some cases, are counterproductive'. Reflecting their acceptance of climate change and expectations of regulation, Table 6.1 shows that the majority manage their responses at the level of the Board of Directors, usually in a specialized committee. Most also offer incentives to managers (and sometimes staff) indicating that they believe the challenge requires a cultural change within the firm, or internal operational targets that must be met.

As US managers are regularly rewarded with cash or stock-based compensation for meeting important targets, it is reasonable to assume (1) that firms consider climate-related goals (which as discussed below often are related to energy-savings) to be important; and (2) that monetary and other incentives will encourage managers to meet or exceed internal goals. However, material rewards for action are to the fore in driving change within the companies: 62 percent of sample firms consider climate-related goals to be sufficiently important to offer mon-

Table 6.1 Governance

Description	Percentage of Firms
Governance of climate change at board level	60
Governance of climate change by chief executive	4
Governance of climate change by senior manager	36
Monetary incentives to management	62
Non-monetary incentives to management	51
Both monetary and non-monetary incentives	36

Source: CDP reports. Totals for incentives do not add to 100 percent because some firms offer both monetary and non-monetary incentives and others offer no incentives.

etary incentives, with 58 percent of those firms also offering non-monetary incentives (such as recognition and awards).

Risks

In addition to accepting the climate change science, most of our sample firms also stated that they were strategizing as if regulation to mitigate climate change is likely. Du Pont, for example, comments that 'additional federal and international policies will be implemented in the coming years that drive a carbon price through the economy'. However, they were wary of the potential costs that regulation will impose. For example, Delta Airlines stated that it 'expects that cap and trade schemes such as EU ETS will impose significant costs on its operations'. In this context, it was notable that many companies stated a preference for a 'legislative approach to dealing with this issue rather than regulation' (American Electric Power) because specific regulations may change every four years whereas they desire more predictable *rules*. Rather than costs to be imposed, specific taxes or trading schemes, they expressed a desire for a clear, enforced legal framework to which they could profitably adapt their products and processes. Cummins put the case in the following terms: '[i]f regulations are not clear or do not provide sufficient lead-time, then we may not have products ready to sell in a market'. It is also possible that companies prefer legislation because they are more able to influence Congressional acts than administrative regulations, but the point is that while nearly all the firms considered regulation 'virtually certain' within the next ten years, Table 6.2 shows that they also consider the risks produced by regulatory uncertainty their greatest challenge. On average they also scored this risk as potentially the most damaging.[1]

Aside from increased energy costs from regulations to mitigate climate change, several firms complained about the range of different regulatory jurisdictions to which they are subject and the administrative costs that these impose. For example, Waste Management comments that 'inclusion of landfills in the Cap and Trade Program being implemented under the California Global Warming Solutions Act would add costs to WM's landfill operations in California'. In general, it was noted that the need to measure and report GHG emissions across multiple customer industries and jurisdictions requires an accounting system to assess the 'carbon footprint' of hundreds or thousands of products. Standardization of regulations across industries and markets would encourage more firms to establish the necessary internal systems. As 3M notes, 'one standard, practical approach would be less

Table 6.2 Perceptions of climate change risks

	No. of Firms	Index of Perceived Severity
Risk of regulatory uncertainty	45	1.51
Risk of physical changes	42	1.02
Other risks	15	0.87
Risk of change in consumer behavior	14	1.29
Reputational risk	14	1.14
Risk change in market signals	6	1.17
Risks from socio-economic conditions	5	0.60

Source: CDP reports. The index of perceived severity is the average score across all firms identifying each specific risk (see Appendix 1).

challenging and financially less burdensome'. For MNCs in particular, multiple jurisdictions can add significant administrative and potentially financial costs. Several corporations mention the different Australian, UK, and EU programs for pricing carbon as problematic because they each impose different sets of rules, measures of carbon emissions, and carbon accounting requirements. Others note that they advocate with policymakers for an international solution to 'level the playing field' and simplify administration.[2]

It is notable that, as shown in Table 6.2, firms anticipate relatively lower risks from changes in actual physical conditions as a result of climate change. Electric utilities may face limited availability of water for cooling and steam production, plus weather-driven changes in electricity demand, and a few firms fear more extreme weather events such as tornadoes in the US Midwest. Yet, despite nearly as many firms expecting physical as regulatory changes most expect that the severity of the effect of the former will be significantly less than the latter. Table 6.2 also shows that while many fewer firms identified changes in consumer behavior as a risk, the firms that identified this risk rated it nearly as severe as regulatory uncertainty and as much more challenging than actual changes in climate. Putting it simply, risks to their financial performance, particularly from regulation or changes in consumer behavior, are rated much more highly than the physical risks from climate change. Without stable and predictable regulatory intervention in particular, they will feel less inclined to reduce their GHG emissions.

Table 6.3 Perceptions of climate change opportunities

	No. of Firms	Index of Severity
Opportunities from regulatory changes	39	1.46
Opportunities from physical changes	31	0.90
Opportunities for enhanced reputation	14	1.29
Opportunities for new products or markets	1	3.00
Opportunities from customer behavior change	23	1.26
Opportunities in market signals	1	2.00
Opportunities from changed socio-economic conditions	6	1.67
Opportunities – other	16	1.19

Source: CDP reports. The index of perceived severity is the average score across all firms identifying each specific risk (see Appendix 1).

Opportunities

Table 6.3 shows that many firms see opportunities in climate change regulation. In fact, regulatory changes are cited more than any of the other climate change opportunities. Thus, uncertainty about the shape of a regulatory framework generates both risks and opportunities, often within a single company. For example, Dow Chemical cites several examples of how regulations may increase operational costs but it also believes that regulations may generate greater demand for current products and open opportunities for new products. Regulations to reduce GHG emissions inevitably will produce both winners and losers. Any regulation that puts a price on carbon emissions would increase business opportunities for firms whose products reduce energy use or enable renewable energy production, or which anticipate regulations to better exploit the economic conditions they create. For example, Applied Materials, a manufacturer of batteries and the equipment used to make computers, cell phones, and solar panels, believes its early actions to 'address climate change and opportunities give the company a competitive advantage'.

Similarly, changes in consumer behavior from climate change also present some of the best opportunities as well as risks. For example, if consumers begin to demand more energy efficient products or cleaner production techniques, firms that already have those products would likely benefit. As Honeywell put it, 'nearly 50 per cent of Honeywell's product portfolio is linked to energy efficiency. The United States could reduce its energy consumption 20 to 25 per cent by immediately and comprehensively adopting existing Honeywell technologies.' Of course, while full diffusion of available and near-production energy saving technologies may potentially reduce GHG emissions by 50 percent, this does not explain why products embodying these have not already been adopted (Mikler and Harrison, 2012).

Some companies identify opportunities from changes in consumer demand as a result of some possible changes in the regulatory framework. For example, US Steel comments that 'any product regulatory requirements or any other driver that encourages increased product energy efficiencies and/or the development of new technologies will also increase the need for high performance steels and increase our market share for value added products'. General Motors is even clearer about its need to remain market-directed to exploit opportunities from regulatory changes: 'customers' needs will remain as the central part of our development process, and we are committed to sustainable transportation that fits their various needs and lifestyles'. Like these two examples, it is probable that many companies anticipate they will be able to adapt to regulatory demands by satisfying the resultant changes in customer demands.

Other firms stated their belief that the GHG benefits of their products may offset the GHG costs to their operations. For example, Dow Chemical claims that the use of some of its products saves as much energy as is used in their manufacture: 'the greenhouse gas emissions avoided by the use of many Dow products (STYROFOAM™ brand insulation, for example) are estimated to be larger than Dow's total direct emissions of greenhouse gases, giving the company a "negative" overall footprint'. However, Dow's direct and indirect emissions are large: '42 per cent of Dow's 2010 Production Costs and Operating Expenses were related to hydrocarbon feedstocks and energy.... In 2011 Dow spent approximately $2.7 billion to purchase energy (fuel, electricity and steam).'

Firms in the same industry may perceive the risks and opportunities from climate change quite differently. For example, General Motors sees high risks from changes in consumer behavior while Ford only sees opportunities; Dow Chemical expects both risks and opportunities of uncertain magnitude while PPG sees only substantial opportunities; General Electric suspects there may be opportunities from consumer changes but cannot quantify them; but much smaller Parker Hannifin sees only risks even though 'energy resources conservation' is part of its 'core business strategy'. The regulation of climate change may cost Parker Hannifin 'about 10 per cent of [their] annual sales or

[$]1.2 billion over the next decade'. In contrast, Johnson Controls is 'experiencing many new opportunities as consumer attitudes and demand continue to focus on energy efficiency'. Pall Corporation that 'solves complex filtration, separation, purification and contamination control problems...has developed a strategy to exploit the opportunities climate change provides for expanding our market share'. Yet, it expects to exploit these opportunities 'with little additional investment'.

In summary, both risks and opportunities remain largely subjective and quite uncertain. But they are cast mainly in light of regulatory and market, rather than physical, changes. Companies recognize that changes in regulations produce both risks and opportunities and whether a particular style of regulation is beneficial or costly is determined not only by their condition – the industry they are in, their current energy efficiency, and so on – but also on their creativity in identifying and exploiting market opportunities.

Long-term versus short-term investment

The CDP questionnaire asks respondents to describe projects implemented in 2011 and to give their expected payback period. One of the choices of payback period is 'greater than three years'. We expect that firms with a high proportion of climate innovation projects over this longer period have a longer-term strategic horizon, and therefore a greater propensity for radical innovation. This is because a willingness to invest for a longer period in an uncertain regulatory environment and accept a lower internal rate of return suggests that the firm is factoring more than immediate financial targets into their investment decisions.[3] Across the 42 firms that report the relevant data, on average one in three climate projects have paybacks exceeding three years. Yet only eighteen firms have a ratio of longer-term paybacks projects to shorter-term projects equal to or larger than 33 percent.

As Table 6.4 shows, most firms (nearly 96 percent) report that climate change presents risks that require assessment and management and that about 94 percent have integrated consideration of climate change into their corporate strategy. However, only half of the sample firms include specific comments on how climate change potentially affects their strategy either in the short- or long-term. This proportion roughly correlates with the ratio of longer versus shorter-term payback climate projects.

Only two firms clearly stated long-term emissions targets: Dow Chemical and Ford. Dow pledges that 'scope 1 and scope 2 Kyoto GHG emissions will not exceed 1990 levels through 2025'.[4] In other words,

Table 6.4 Firms including climate change in risk assessments or strategy

Perception	Percentage of Firms
Climate change integrated into risk management process	96
Climate change integrated into strategy	94
Climate change affecting short-term strategy	49
Climate change affecting long-term strategy	51

Source: CDP reports. Includes all firms in the sample. CDP respondents were encouraged to specify how climate change impacted short- or long-term strategy but only about half did.

for a dozen years energy intensity will fall sufficiently to offset revenue growth. However, not only does this mean that its absolute emissions are not projected to fall but the exclusion of scope 3 emissions means that the firm's climate 'footprint' could well continue to rise. Ford 'adopted a goal to reduce [their] facility carbon dioxide emissions by 30 per cent by 2025 on a per-vehicle basis' which also is not an absolute target that contributes to the national or global emissions reductions necessary to mitigate dangerous climate change. As they baldly state: 'we expect total emissions to increase due to increased production'. Ford's goal 'is to provide consumers with a range of different options that improve fuel economy and overall sustainability while still meeting individual driving needs' which 'has direct implications for our sales volumes and market share, both of which contribute significantly to the Company's overall financial performance'. This suggests that they remain primarily market-focused and will only increase the energy efficiency of their products as they perceive consumer preferences change in that direction. Economic and financial imperatives trump long-term emission reductions.

We argued in Chapter 1 that climate innovation needs to be primarily about radical innovations that are longer-term in gestation, demonstration, and marketing than incremental projects that are more attuned to market signals. Most of the R&D investments reported by firms were short-term in nature, and primarily in respect of energy reduction to increase operational energy efficiency. This is easier than long-term strategies which, as Deere notes, involve '[i]nstitutionalizing sustainable solutions into our product design process'. In this light, a preference for energy intensity targets (see Table 6.5 below) indicates that companies expect to reach long-term goals through a progression of short-term internal process changes that incrementally reduce the energy content of operations, products, and services. This is probably a reflection of firms' uncertainty about the nature and timing of regulation indicated by the consensus that regulatory uncertainty is the most significant climate risk they face – that is, they expect to progressively adapt to regulatory changes as they occur. For example, Anadarko Petroleum comments that 'climate change has only affected the long-term strategy until recently [sic]. Pending regulations impacting the oil and natural gas industry prompt us to develop short-term strategies to manage these risks and mitigate impacts to operations.' The result is that many firms would likely agree with Krueger International that long-term strategy is based on 'a continued reduction of energy and GHG while considering other options such as renewable energy and more energy efficient manufacturing processes'.

Energy intensity and process changes

As Table 6.5 shows, our sample firms expressed a preference for energy intensity targets that usually are pursued with projects to increase energy-efficiency within internal operations. A smaller group of firms have both an absolute and energy intensity target. But only two firms (Dow Chemical and Ford) clearly stated any target beyond 2020, the target year for the first stage of US national emissions reductions. General Electric implied that its 'Ecomagination' initiative generates products that could reduce emissions beyond 2020 – as did a few other firms – but does not state a longer-term target. Eight firms (predominantly in resource extraction) have not adopted any target for their GHG emissions arguing, as does mining company Freeport-McMoran, that 'changing market conditions' prevent meeting targets: 'direct and indirect emissions are directly related to changes in our mining production, which is correlated to global economic conditions'.

Table 6.5 Percentage of targets by type

Target	Percentage of Firms
Absolute emissions	26
Energy intensity	36
Absolute and intensity	19
No target	19
Target beyond 2020	4

Source: CDP reports. The percentages add to more than 100 per cent because the firms with targets beyond 2020 are included in percentages above.

Because intensity targets are indexed to some financial or operational measure that is expected to grow, such as revenue or number of employees, they allow firms to look green while not achieving the absolute emission reduction goals required to mitigate dangerous climate change. Firms use several justifications for preferring them to absolute targets. For example, Eastman Chemical says: 'we focus on reducing the energy intensity of our products rather than reducing absolute emissions, as greater use of our products actually results in a greater net reduction of emissions in the total economy'. This presumes growth in the indexing measure but also ignores the need to accelerate absolute emissions reductions from historical levels.

In several instances, firms report their recent successes in reducing energy intensity without offering a new target beyond 2011. For example, although 3M 'has been setting environmental goals since 1990' it pursued an energy intensity target which expired in 2011 which has not been formally replaced. However, it did achieve its most recent target, an aggressive 55 percent energy intensity reduction between 2006 and 2011 and a 72 percent *absolute* emissions reduction since 1990.

A minority of the sample firms published both absolute and intensity targets. Table 6.6 shows the range of targets by approximate annual emissions percentage reductions.[5] This distribution also shows the preference for intensity targets. But it also shows that whether the firm has a single target or dual targets the annual reduction in intensity targets is slightly more aggressive than for absolute targets. This is to be expected as the same annual reduction rate amounts to a greater reduction in GHG emissions for absolute targets than for intensity targets.

Table 6.6 Number of firms declaring annual absolute and intensity targets

	Single Target		Dual Targets	
	Absolute Target	Intensity Target	Absolute Target	Intensity Target
0			1	2
<1%		3	2	
1–2%	4	4		
2–3%	1	4	2	6
bg3%	3	1		2

Source: CDP reports. Only includes firms that clearly indicate their target emission reductions, and includes firms that cite targets expiring in 2011 (see Appendix 1).

Assuming growth occurs, a 3 percent reduction in energy intensity is likely to result in an absolute reduction of emissions of less than 2 percent. And reducing energy intensity is primarily about energy saving. For example, US Steel makes no bones about its priorities stating that their 'targets are strictly energy reduction targets that [they] expect will also result in CO_2 emission reductions'.

We may speculate that once climate change regulations increase the relative cost of fossil versus renewable energy, more firms are likely to switch to absolute emissions targets. In fact, companies that consume large amounts of energy in their production processes and that are willing to invest in advance of emissions regulations are constructing their own renewable energy sources. Two examples are Dow Chemical and Alcoa that are building hydro-electric plants. It is quite likely that they are doing this both to reduce the risk of price rises in essential energy inputs over the long-term and also to prepare for regulations that they expect to put a price on carbon and drive up their fossil energy costs. Projects such as these which are designed to meet absolute targets have a longer time horizon, but for now, most firms are primarily focused on short-term intensity projects that reduce fossil energy consumption.

Low hanging (economic) fruit

Recognizing that the CDP reports are prepared for the institutional investor community, it is no surprise that statements such as this from Apache Corporation are common: 'we focus our efforts on where we can gain the greatest return on our investment of money and time in reducing GHG's emissions.... Our strategic goal is to maximize value for our shareholders'. Similarly, Eastman Chemical states that it has 'identified thousands of projects to pursue and are approaching them on the basis of highest return on investment and availability of resources'. Conoco Phillips says that 'the primary goal of [their] Corporate Climate Change Action Plan is to prepare the company to succeed in a world challenged to reduce GHG emissions. "Success" is satisfactory shareholder return while operating in an environmentally and socially sustainable manner.' Clearly, economic and financial goals are considered more important than climatic ones.

To ensure the priority of investor returns, most firms have focused on 'plucking' low hanging fruit. The attractions are obvious. As Schlumberger notes, 'due to the high costs associated with operating our equipment, actions taken to reduce emissions often result in cost benefits and efficiency improvements in operations'. In other words, the key reason companies are chasing emissions reductions in the form of energy savings is not just that these are short-term and less risky, but that they contribute most immediately to profits. This focus is economically rational when the cost savings per tonne of CO_2 are high. One project implemented by Praxair, for example, generated '$62 million of savings in 2011, and 306,000 MT CO2e [sic] avoided'. This equates to earning $203 per tonne of emissions avoided which is much more attractive than selling tradeable permits under EU ETS for $50 – the price of permits in 2006, and from which point it has continually declined.

Several of the firms also recognized economic advantages from selective investment in future technologies. Cummins, a manufacturer of heavy equipment (usually diesel) engines, expresses the belief that 'sustainability starts with a strong financial performance', but also expresses an expectation of financial benefits from climate-related fuel efficiency research which it said will increase from 10 to 20 percent of its R&D expenditure over the next five years.

The theme of increased energy efficiency as the driver of innovation was nearly ubiquitous. For example, Du Pont claims to have saved $200 million through its energy efficiency program, because it sees energy efficiency as a 'key short- and long-term climate mitigation strategy'. Because fuel is their primary cost, airlines compete on fuel-efficiency regardless of any impact on GHG emissions. Southwest Airlines has a young fleet so they 'could have a competitive advantage over airlines with older, less fuel efficient fleets if energy taxes or regulations were enacted'. Although United Continental says it is 'committed to leading commercial aviation as an environmentally responsible company', its primary commitment is to 'reduce fuel use and improve fuel efficiency of (its) aircraft and operational procedures through technology and process innovation'. Similarly, electric utility companies focus on energy savings within their operations because energy is such a large proportion of their costs – between 45 and over 90 percent depending on the technology and fuels used in their power plants. The long lead times for investments in new generating equipment has focused their attention on shorter-term projects designed to wring fossil energy out of many aspects of their operations. However, for the electric utilities in our sample nearly 60 percent of their emissions reductions projects had a payback period of more than three years.

The firms also include in their reports actions that may reduce emissions in their processes or their customers' operations as a result of their products. Sometimes these emissions reductions are incidental or again clearly driven by financial motives. For example, the oil and gas production companies claim a switch from oil production to natural gas as a primary contributor to their emissions target. Anadarko explains that its business strategy involves 'the promotion of and increased production of natural gas as a market commodity and alternative to carbon-intensive coal'. Similarly, Noble Energy and Exxon are able to count emissions reductions against their internal targets by following an industry trend and investing in gas production. The impetus of this change is the use of innovative 'fracking' technologies that initially were focused on 'tight' or shale gas formations. Because the majority of oil production is now owned by national champion corporations like Pemex in Mexico or Aramco in Saudi Arabia, domestic gas fields in the US are a primary option for exploration and development companies.

In the short-term, such innovations as the substitution of gas for coal are beneficial both economically and environmentally. This process initiated by the abundance of gas (from fracking) and underpinned by economic considerations would likely be hastened by the regulations that the Obama Administration has proposed.[6] However, the burning of gas still produces GHGs (though fewer than coal) and, compared to the development of more radical renewable energy sources, slows progress towards the 80 percent emissions reductions required by the US by 2050.

Climate innovation and diversification

If US firms continue to prioritize economic considerations over GHG emissions reductions, a price on carbon to increase the cost of fossil energy would seem to be one of the more effective policies. The downside of such an approach could be reduced economic growth. Although this also would tend to reduce fossil energy consumption (as it did in 2008 after the global financial crisis), there would also be a cost to social welfare. Climate innovation that is predicated on an aggressive search for technological alternatives should be both more acceptable to firms and contribute to economic growth.

However, given the scale of the task to reach even the US's modest GHG emission reductions by 2020, it is discouraging that reductions in operational energy intensity remain the primary motivation and that development of new and radical technologies is less stressed. Only a few of the firms expressed a more systematic approach to climate change. For example, United Technologies, a diversified defense and aerospace firm, accepted that its business strategy had been influenced in three main areas: energy management to reduce GHG emissions at its production sites; product research and development; and strategic partnerships with suppliers to increase their efficiency. General Motors claims the following:

> [we are] on a journey to reinvent the automotive DNA, which is driving a great amount of innovation and technological breakthroughs. We have an aggressive focus on advanced propulsion technologies that will benefit customers and the environment. We focus on inventions that make our vehicles more sustainable.

As a result it has won awards as the 'No. 1 innovator' for several quarters and 'during the past 10 years, it increased its patent filings sixfold'.

General Electric too has identified energy efficiency as a primary driver of company financial performance. By 2012 it had achieved its goal (originally set for 2015) of reducing energy intensity (indexed to revenues) by 50 percent from 2004 levels. This is equivalent to about a 25 percent reduction in absolute emissions. In part it achieved this by recognizing that firms 'in the business of innovation and technology that embrace this opportunity [climate change] will lead and win'. It further supports the case for climate innovation by explaining its success:

> To accelerate innovation we committed to double our annual investment in clean tech R&D by $1.5Billion by 2010. We accomplished that in 2009, a year ahead of plan, delivering even through a severe economic downturn. This accelerated R&D investment totaled $5 billion and produced a fivefold increase in certified products since 2005.... The ecomagination portfolio currently includes more than 140 products and solutions. We have generated more than $100 billion in revenue to date, exceeding our ecomagination growth targets and growing at a faster rate that the full GE portfolio.

It expects that by 2015 it will 'double R&D to $10B; Grow revenues from eco-certified products at a rate 2X GE's growth and Reduce GE's energy intensity by 50 per cent'. In a similar vein, 3M, another diversified manufacturer, is investing in 'initiatives to identify and drive the utilization of renewable materials as replacements for petroleum based' raw materials and estimation of energy consumption in product life cycle analyses.

Dow Chemical also claims some substantial climate innovations from 'increased R&D funding for clean energy technologies' such that some 20 percent of its R&D Budget is spent on the development of clean energy technologies. In addition to Styrofoam insulation:

> Dow POWERHOUSE™ Solar Shingle integrates low-cost, thin-film CIGS photovoltaic cells into a proprietary roofing shingle design, which represents a multi-functional solar energy generating roofing product [and] has been hailed as revolutionary, including being named one of the '50 Best Inventions of 2009' by TIME magazine and winning the GLOBE Award for Environmental Excellence in Emerging Technology.

Other Dow technologies appear more incremental such as 'prismatic lithium-ion batteries...store up to three times more energy' than current hybrid car batteries; epoxies to strengthen wind blades and lower their weight; ENLIGHT™ SilverPlate 620 technology that 'increases solar cell efficiency'; improved structural adhesives to reduce automobile weight and reduce costs; or roof coatings to reduce solar gain. Du Pont – in many of the same markets as Dow – spends 86 percent of its $2 billion annual R&D budget on 'feeding the world, decreasing our dependence on fossil fuels, and protecting people and the environment'. However, the majority (62 percent) of this R&D effort is related to food production through its genetic management of seeds.

Air Products states that it 'is investing significantly in research and development of offerings that enable its customers to reduce their environmental footprint and energy consumption'. It spent about 60 percent of its $119 million 2011 R&D budget on projects related to environmental and energy efficiency solutions such as 'low carbon product and processes'. Its competitor, Praxair, sees 'long-term business opportunity from innovation that takes advantage of opportunities presented by climate change'. However, its long-term target (to 2020) is focused around saving energy in production and operations 'delivering anticipated savings in excess of $600 million and 6 million mt [sic] CO2e by the end of the goal period'. Again the financial consequence of internal process enhancements appears to be the primary motivation. It offers few details on its R&D but does claim that the list of projects under review 'contains a potential 2 million MT of GHG avoided'.

There are other examples of corporations undertaking more aggressive climate innovation. However, what this entails takes on different meanings depending on the industrial sector. In particular, it is a greater challenge for those industries with high and long-term capital needs. For example, airlines and electric utilities have capital plans that are predicated on large, long-term capital expenditures to meet forecast demand for their products and services. Airlines buy or lease their aircraft to meet predicted passenger volumes up to decades in advance. Regulatory or market changes only marginally impact their purchasing plans, unless they anticipate such future changes and invest in more fuel and emissions efficient aircraft now. Even so, in reality Delta reports that 'to achieve the short-term industry fuel efficiency goal of a 1.5 per cent annual improvement in fuel efficiency through 2020, 12,000 new aircraft will have to enter service between 2010 and 2020, at a cost of $1.3 trillion to airlines'. Furthermore, their strategies are constrained by the initiatives taken by the main passenger aircraft manufacturers and the aircraft they offer.

Because power plants take several years to design, be approved, and built, and usually operate for over fifty years, electric utilities have to predict both consumer demand and environmental regulations decades ahead to plan their generating capacity. Regulatory uncertainty encourages utility firms to prefer conservative strategies, such as demand reduction through multiple small projects from energy audits to smart meters. Exelon, for example, states that it 'has a tremendous opportunity to assist its customers in managing electricity consumption, needs and costs, and overall improving the long-term business of electricity transmission and distribution'. But it and Duke have advantages over others such as American Electric Power which has more coal-fired stations, and therefore fewer options in this regard. In addition, all electric utilities are at least partially regulated by the states in which they operate, creating further investment rigidities, and none is able to diversify its products or easily expand geographically.

Finally, it appears that more diversified corporations, with multiple business units, have greater flexibility in their climate innovation strategies and more opportunity to profit from regulatory and behavioral changes. As Waste Management comments: 'Our CEO has set and our Board has approved aggressive sustainability goals with ambitious emissions reductions benefits. Moreover, there is no limit to the number of emissions reduction activities available to a highly diversified company like WM'. Diversified chemical and industrial firms similarly are able to adapt their products for a climate constrained world either to mitigate or help society adapt to climate change. It is also the case that corporations focused on a single business area but geographically diversified enjoy advantages in adapting to climate change regulations. Alcoa, a global aluminum producer, believes that its strategy of 'aggressive targets, inert anode R&D, hydro-electric-based investments in Iceland, Brazil and Quebec, new integrated facilities in Saudi Arabia, product innovations and recycling alliances will give it a leading climate advantage in the aluminum industry'. Thus, MNCs may have an inherent capacity to generate cost-effective innovations from effective adaptive responses to opportunities in multiple jurisdictions that are not replicated by smaller firms with wholly US operations.

Australian corporate perspectives

In late 2012, we interviewed twenty senior office-holders from fourteen leading Australian corporations. The firms were drawn from the following Dow Jones Sustainability Index (DJSI) industry sectors: airlines; building materials and fixtures; heavy construction; electricity; mining; oil equipment and services; oil and gas producers; waste and disposal services; steel; and industrial transportation. These sectors were selected on the basis that they are major contributors to GHG emissions, and are therefore crucial for climate innovation. Those interviewed included managers of innovation, sustainability, public policy, marketing and product design. The intention in interviewing them was to get first-hand perspectives from those in high-level strategic positions in the immediate aftermath of Australia's implementation of a $23 per tonne tax on GHG emissions for the 2012/13 financial year, as opposed to focusing on official corporate policy statements which would not have captured the impact of this event. All interviews were also conducted with assurances that both the interviewees and their companies would remain anonymous, with the result being that the views expressed are interviewees' personal opinions, not the official statements of the corporations for which they work. The questions asked ranged from the general nature of the climate innovation being undertaking, before turning to the specific government policy, market and internal corporate drivers (see Appendix 2).

As shall become apparent, the views expressed are strikingly similar to our findings in respect of US CDP reporting firms.

The climate innovation being undertaken

Climate change was seen by all interviewees as one of the greatest challenges their companies faced. They all stressed that they accepted the scientific evidence on climate change, with the level of 'fear mongering' around climate skepticism seen as something both regrettable and unusual to Australia. As one interviewee put it, 'in Australia we've had this extraordinary doubt-the-science movement. If you look back to Australia from Europe or America or other places, it seems quite absurd.' It was in this context that they discussed their companies' climate innovations. They were asked whether these were more radical or incremental, and invited to define what might be meant by this. They primarily defined them in terms of the extent of GHG emission reductions. Incremental innovations, including the application of existing technologies, were seen as producing up to 5 percent reductions, while more radical innovations involving a technological 'step change' were seen as those that reduced GHG emissions by magnitudes of 10–20 percent or more. Anything higher than this was described by one interviewee as 'really leading, breakthrough-type stuff'. It is interesting to note that the 5 percent annual reductions threshold beyond which innovations become radical is higher than nearly all the US sample firms have been able to achieve.

Importantly though, to some extent all innovations were seen as being radical in their impact on *corporate practices*. This is because even incremental changes involve initial costs, and what might be described as an upheaval to business operations. Putting it bluntly, one interviewee declared 'we just think it's bloody hard work!' Even 1 percent incremental efficiency improvements on a year-by-year basis that could add up to substantial savings over time through redesigning processes and procedures were perceived as involving a comprehensive and major effort on the part of the company. They were therefore 'radical in decision making and incremental in terms of technology deployment'. As such, what might be regarded as marginally beneficial initiatives by an outside observer, such as assessing existing processes to see what technological 'add-ons' were possible, or redesigning existing products, were regarded as requiring a major firm-wide strategic commitment not entered into lightly, especially in times of uncertain economic returns. As another interviewee put it, 'if you understood the effort that went into gaining what you term as an incremental change you'd be amazed'.

It followed that radical technological innovations that had the potential for the biggest impact were the least readily embraced. This is because they involve substantial capital expenditure, and major changes to business processes and products for uncertain future benefits over long periods of time, with great potential for significant losses. Indeed, several interviewees cited losses in the hundreds of millions of dollars on more radical technological research and development they had previously undertaken. Therefore, the business realities actively discourage radical climate innovation. This was why one interviewee said, 'we're only really interested in incremental (improvements) because they are the only things that work', before going on to comment that 'you can't make any money out of radical'.

All interviewees therefore stressed that any climate innovation had to be undertaken with the 'business case' for it in mind. Increases in efficiency, the likelihood of productivity gains and/or cost savings that flowed from these were the main drivers. The extent to which GHG emission reductions resulted was said to be determined not by a desire for this, so much they were the consequence of motivations for positive financial outcomes. The corporate realities meant climate change mitigation could never be the primary reason for investing in technological innovation, even if there was a desire that this be one of the outcomes. As one interviewee put it, 'there's not a lot of point spending a lot of money to save the planet if the projects aren't going to work'. This sounds like an odd statement, as the destruction of the planet would surely be bad for business. Even so, the point is clear.

As there always must be a strong business case for investment decisions, justifying technological innovation on the basis of climate change mitigation alone was never enough for approval to be granted. An economic and environmental 'double dividend' had to be demonstrated. We noted above that many US firms reported a priority for economic or financial justifications for innovation investments and that emissions reductions are seen as an attractive by-product. The Australian interviewees essentially concurred that the business case has to be made before the environmental benefits are considered.

The role of government

Given the lack of a business case for climate innovation purely on the basis of GHG emission reductions, all interviewees stressed that those of the radical variety in particular, entailing the entire redesign of processes or products, were highly unlikely without strong regulatory requirements and substantial support on the part of government. To one degree or another, they echoed the sentiment that 'regulatory is by far the strongest driver' of GHG emission reductions and that 'regulations drive innovation'. The Australian interviewees echoed the US firms' formal statements of position: regulation is the most powerful incentive for firms to innovate to reduce GHG emissions, and given the scientific evidence, regulation is necessary and expected.

In this context, it was interesting to note that none of the interviewees were opposed to putting a price on carbon, with the majority saying this was important. All of them also perceived the introduction of a carbon tax in Australia as symbolic, and perhaps a harbinger of future enhanced regulatory requirements for which they all saw a need. Echoing the point made earlier about price inelasticity of demand, it was interesting to note that nearly all the interviewees said the tax needed to be much higher, with one commenting that although it was high enough to reduce profitability it was not high enough to substantively drive innovation. One interviewee put the case thus: 'either you put it in as a token leadership issue, and awareness issue at something like $10 a tonne which we could easily cope with, or you put it in at $60 or $70 which would actually drive innovation and change'. None of them saw this as politically feasible though.

The main difference of opinion on the carbon tax was on the extent to which they saw it as impacting their ability to invest in the necessary technological innovations. This largely turned on the extent to which they were trade exposed. Those that were not trade exposed said the carbon tax had been very important in making a business case for deploying new innovations. It had enhanced the business case for reducing GHG emissions, and as they were able to pass through the cost of the tax to their consumers, it also raised awareness and acceptance in the market of the benefits. While interviewees from trade exposed corporations agreed that the tax needed to be much higher to have a real effect, this did not mean that they embraced the prospect. At present, they saw the tax as a measure they could cope with because they are allocated emission permits in recognition of their trade exposure. However, as such arrangements were wound back, they saw themselves in a much more precarious position as they could potentially be driven out of business by foreign competitors. In the absence of a global agreement on pricing GHG emissions and/or a global carbon emissions trading scheme, the result could simply be the off-shoring of emissions to countries without such measures. This was seen as completely counter-productive.

Even worse, being unable to pass the costs on to consumers, in the shorter term they found that the uncertainty surrounding their companies' futures meant they were actually *less* inclined to invest in climate innovation. Rather than producing a business case for so doing, they saw incentives to 'just say for the next four or five years we will defer this'. They faced a 'Catch 22' situation: while arrangements stood as they were there were economic and policy incentives for the status quo, but the forecast change in these arrangements would simply have economic impacts without the climate mitigation benefits. As one interviewee put it, 'all that happens is that emissions get produced in some other country'.

Another point made by all interviewees was that much more was required of government to drive climate innovation, especially that of the more radical variety. They all bemoaned programs of a limited duration that followed political cycles and subsidized certain technologies, with the subsidies then removed within a few years. These just produced short-term 'band-wagoning', as companies made what profits they could and consumers acted self-interestedly to accrue the financial benefits while the programs lasted. One interviewee commented that in a policy sense 'at the moment it feels a little like mud's being splattered against the wall and some of it sticks and some of it doesn't'. To produce real innovation that led to sustained change, a stronger and longer-term strategic commitment on the part of government was required, as well as consistency across the many arms and levels of government. As one interviewee put it, 'climate innovation has got to be long term, so there's got to be a strategy and it's not about short-term programs'. Given that more radical climate innovation involves substantial capital expenditure and a five to eight year commitment at least, with the prospect of uncertain future returns over a longer period of time after this, 'if your legislation is changing on a six monthly basis you just can't do it'.

In addition to market mechanisms such as the carbon tax, all interviewees therefore said that much more extensive regulatory standards were necessary – for example, strict efficiency and GHG emission standards with which all competitors, whether nationally based or located overseas, would be compelled to comply as part of the (non-market) 'price' for doing business in Australia. This did not mean picking certain technologies, or expanding the plethora of standards and requirements already in existence. It meant streamlining these, setting clear emission reduction targets coupled with product and process standards, and then allowing corporations the freedom to meet these as the market dictated – that is, setting new rules for market competition.

All interviewees also said much higher funding for innovation was required, given the expense and uncertainty associated with it, and that the system for accessing the funds needed to be clearer. The complexity of the policy environment in respect of funding climate innovation, coupled with the many programs and varying requirements at different levels of government, was also cited as a problem in this context. One interviewee literally threw up his hands and exclaimed 'I just can't see how a small to medium company could ever negotiate this process and ever get money out of government. It's just so difficult!'

While the need for longer-term policies, better coordination between levels of government, easier to access funding schemes and clearer regulatory settings to support climate innovation were widely discussed, unfortunately no interviewees expected this. A lack of bi-partisan political support, and the greater potential for more rapid short-term policy changes as a result, meant that as one of them put it, 'the politics has overrun the strategic thinking'. The result was said to be that 'people aren't investing their money in (GHG emission reduction) projects because we don't know what the policy framework will be'.

The role of the market

All interviewees stated, often quite bluntly, that they perceived *no* business case for climate innovation specifically. This is because they did not believe consumers were sufficiently demanding less GHG emissions intensive products, unless they can be provided at the same or lower cost. With the costs and risks involved in climate innovation for such products, there was therefore limited incentive to invest in them. As one interviewee said, 'the options around the consumer driving it are fairly limited', while another noted that 'we can't build a model around…the top two per cent of consumers who will buy green products'. As such, another stated that 'if you let the market do what it wants hardly anything will happen. They'll just go to the lowest cost solution, which is the least environmentally friendly solution.'

Similarly, shareholders were not seen as driving the investment in new products or new production processes that are less carbon intensive if these result in lower profitability and smaller dividend payments. This was why one interviewee declared that 'the biggest driver for a business is shareholders. What keeps us going are our shareholder returns, so that has to be the underlying driver'. Furthermore, acting contrary to their fiduciary responsibilities to shareholders for the sake of mitigating GHG emissions could incur legal action in addition to shareholders' wrath. Shareholder returns were said to therefore be 'the underlying driver' of investment decisions. As another interviewee put it, 'whilst there is lots of lip service being given by companies to being good corporate citizens, basically the major priority is providing shareholder value and therefore things need to be economically based'. This stands in stark contrast to the very impetus behind the CDP, which is based on what would seem to be the somewhat dubious premise that institutional investors will apply the necessary pressure.

As such, although there were the usual business drivers to enhance their corporations' product and process cost, efficiency and productivity profiles, primarily by reducing energy use, the market drivers for climate innovation *specifically* were harder to discern. Although some discussed the impact of industry standards such as efficiency ratings that influenced their corporation's competitive standing in certain markets, what was much more strongly emphasized was the importance of a market case being created by government through subsidies for research and development, or for the deployment of certain technologies in products to consumers (for example, rebates for home insulation and solar hot water heating), or regulatory targets that required compliance. *In fact, it was striking that when asked about market forces, all interviewees talked about government intervention.* The sentiment of all interviewees was summed up by one who said 'if it's not supported by government, then they vote with their feet, the public vote with their feet, and whatever's most cost efficient they'll move to'. In the absence of this support, none of the interviewees saw market imperatives for climate innovation, either now or in the future, despite raised awareness. Indeed, one said that 'what the community expects is that government will reflect their attitudes because they're not going to pay for companies to reflect it', while in a similar vein another said that 'you can't afford to create awareness. It costs too much money. The very best way of initiating change is through government regulations'.

To the extent that they perceived incentives for climate innovation, as opposed to normal market and shareholder-led innovation, in the absence of government intervention they saw this as being driven by a need to reduce risks and address the physical effects of climate change that they thought likely to impact on their profitability *in the future*. As one interviewee put it, 'from an adaptation point of view it's risk management 101, and we have all those processes in place, they just get amplified'. Another said the following: 'if we are going to have a country that's hot or drier or more prone to cyclonic events, that really is an important driver for us in our innovation space to ensure that we've got the products that meet those emerging needs'. In other words, there were forecast market opportunities and risk management imperatives in innovation for climate change adaptation, while there are more moderate efficiency drivers in innovation for climate change mitigation.

The responses of the interviewees therefore lend support to the paradox we have previously suggested in respect of a liberal economic preference for markets as coordinators of economic activity, with this in a specifically Australian context. In 'private', these corporate leaders all expressed a need for more, not less, government intervention in advance of a crisis. To the extent they did not say this in public, this was due to the absence of a stable long-term policy environment, and the politically charged nature of the debate that they did not think would improve. However, they would accept *greater* state intervention that built a business case through increased public investment in innovation; incorporated strict across-the-board regulatory standards that applied to all corporations, whether domestic or foreign; educated the market; and that involved a comprehensive and long-term strategic public policy approach across all levels of government – that is, 'a single national carbon policy'. An absence of this type of policy environment for climate innovation was actively producing either inactivity, or conservative choices of those technological solutions that were less radical. One interview explicitly said that 'because of the uncertainty around policy, we're definitely getting sub-optimal outcomes in terms of cost and carbon emissions, but what can we do?' The lack of clear market drivers, coupled with the non-conducive state of the Australian political landscape, encouraged them to invest less in climate innovation, and to implement existing technologies that would reduce GHG emissions by less than was possible.

Internal commitment

Based on the above discussion, it is clear that the internal commitment to mitigating climate change was said by all interviewees to be predicated on the existence of a business case, with this largely driven by government policy. It was a matter of degree though. At one end of the scale were interviewees who dismissed the possibility that there was *any* corporate commitment to mitigating climate change in the absence of this, such as one interviewee who said 'there is no altruism in a company', and who declared 'I don't see any corporate desire to act on climate change'. He saw more willingness for 'greenwashing', maintaining the status quo and marketing efforts by corporations to undermine weak and/or unpredictable government policies because at the end of the day 'they're profit generating mechanisms'. While all other interviewees also made comments along the lines that their overriding aim was 'to deliver the best long-term growth to the shareholders', and none of them saw the potential for 'green philanthropy', most were more pragmatic about incorporating mitigating climate change with the more traditional business drivers. As one interviewee said, 'I can use the language of efficiency and improvement and continuous evolution...to get the right result', because 'there's a very clear imperative internally for us to reduce cost, there's a very clear imperative externally for us to reduce emissions, and the two come together quite nicely'. Nevertheless, the notion that mitigating climate change alone would succeed in winning approval for investment at board level was universally dismissed as impractical and naive. Putting it somewhat colorfully, one interviewee summed it up thus: 'something that's "fluffy", that has a "shithouse" business case, is unlikely to get up. You've got to have the package'.

That 'package' was less in evidence for trade exposed corporations. Their interviewees saw their companies as facing a worst-case scenario with cost and profitability pressures, in a context of current and future policy uncertainty and complexity, and with the likelihood of further future policy changes, yet facing insufficient support for innovation. As noted above, for companies in such a position there was actually said to be an incentive for focusing *more* on the bottom line and less on climate mitigation as a matter of necessity, and also as a result of current compensation measures under the carbon tax. As one interviewee from a corporation in this position put it, 'you get focused on survival, you get short term, you have to, you can't implement anything if you're not here'. The carbon tax as introduced,[7] coupled with an uncertain policy and non-conducive political environment, had actually made them *less* likely to embrace climate innovation. The main decision such corporations were now preoccupied with was whether or not to shut down or shift production offshore at some stage, rather than reducing GHG emissions.

More optimistically, some interviewees expressed a hope for a more enabling climate innovation environment in future. There were precedents to suggest this was possible in government support for innovation broadly defined that had paid off in the past in Australia as well as overseas,[8] and that could permit greater internal commitment to investment in new technologies. But more broadly, and in respect of climate innovation specifically, one interviewee said that it was not just a matter of support, but of setting longer-term targets that would allow his company 'to know what the rules of the game are going to be over a longer period of time' thus encouraging it to 'make those investments now'. As such, clear targets with substantial lead times coupled with support for investment would be highly effective. The ideal situation was described by one interviewee in the following terms: 'we're not climate scientists, but it appears to us that if you would want to do something you would be best doing it sooner rather than later, so that it can be a gradual, transitional change rather than a sudden change'. For truly radical innovation in particular another commented that in the absence of this what his company was able to do was take 'small bets' on demonstration projects in the hope that at some stage government might support the viability of the technologies underpinning these projects by driving investment in the development and uptake of the new products and processes enabled by them.

It is to be hoped that this will happen in advance of a crisis. In the absence of such an investment environment being created now, some foresaw the likelihood of major government intervention as the crisis worsened and adaptation was necessary instead of mitigation, but this was the kind of government intervention that was much less preferred as it potentially involved extremely radical policy options – for example, one interviewee foresaw the possibility of the government turning off the electricity for portions of the day, while another envisaged the banning of certain products and processes. For now, even with a strong internal commitment at the highest level of his company by the CEO, one interviewee noted that 'the uncertainty around climate change policy here in Australia has led us to change our decision to one that's a little bit of a bet both ways'.

Conclusion

In this chapter we have compared the perspectives on climate change of leading firms in the US and Australia. The measures we have used are not directly comparable but they tell strikingly similar tales. The US data are drawn from public reports targeted at the investor community. The Australian data is from interviews that record more personal opinions of executives responsible, or senior participants in, formulating and implementing climate change policy within their firms, and not just in the context of likely future regulation but the recent introduction of it. Nevertheless, the data display several points of commonality that more generally indicate public companies' climate innovation strategies in liberal capitalist economies.

The Australian experience suggests that a carbon tax is no 'silver bullet' solution. To reduce GHG emissions, market mechanisms such as those currently in place in Australia will not be sufficient. The interviews clearly show that without consumers demanding climate efficient products or governments intervening more extensively, business sees no profit in climate innovation *per se*. In short, under current conditions there is no business case for products and services that are *specifically* designed to reduce GHG emissions. Consumers may demand goods and services that are cheaper, faster, and smarter, and corporations wish to reduce their costs in the interests of their shareholders. This may reduce GHG emissions but firms generally expect that a focus on emissions reductions as a priority would result in a loss of market share, reduced financial returns and, possibly, bankruptcy. If consumers do not demand products and services embodying the most advanced technological innovations to produce the lowest GHG emissions or if shareholders do not embrace a long-term strategy to develop and market them, firms cannot be blamed for failing to invest in them.[9] Survival demands that firms *must* remain focused on the bottom line for their survival, even though the environment is expected to be much more unforgiving if emissions reductions targets are not met.

US firms are no different. They generally prioritize the economic over the environmental investing in short term, cost saving projects, that yield immediate financial returns and coincidentally reduce GHG emissions. Several firms clearly state that profits and shareholder returns are paramount and that consumer demands and market signals drive their strategies and product development. Other firms only imply these priorities in their governance structures, choices of emissions reductions projects, and their perceptions of risks and opportunities. The consequence is that most US firms reporting to the CDP prefer intensity targets and invest in short-term projects with high internal rates of return that imply a very high price for the carbon emissions saved. Because the CDP aims to use investor leverage to reduce emissions, it is no surprise that respondents highlight their investments in reducing energy-intensity. Saving energy saves costs and increases profits, and is strongly supported by the investor community (Smithwood and Hodum, 2013).

The exceptions to this pattern are few. As reported, a small minority of firms promote their embrace of climate change in their long-term business strategies and appear to be innovating accordingly. Most of these firms are able to monetize their investment in technological innovation through multiple markets, both categorical and geographical. They also have already constructed the accounting and reporting systems that carbon pricing will require. Twenty-two firms in our US sample already participate in at least one emissions trading or carbon tax system though only a minority of them are investing in climate technology for the long-term. The capabilities and experience of these firms are relatively unique within the US so they do not indicate how US firms in general can or would respond to carbon pricing. However, because they operate in so many countries, their desire to seize the opportunities that climate change presents may begin to change organization processes and attitudes across the globe.

The current rate of decarbonization in both the US and Australia is insufficient and firms are reducing their GHG emissions too slowly (PWC Advisory Services, 2012).[10] It appears from their public statements in the US, and the private comments of their interviewees in Australia, that firms are doing too little because they are focused on their bottom lines. As long as firms think of reducing GHG emissions as a secondary, or by-product, of normal business growth decisions that are driven by investment returns and market exigencies, they will fail to meet aggressive targets for absolute emissions reductions necessary to avoid dangerous climate change. In primarily choosing the low-hanging economic fruit of internal process energy savings, and the incremental introduction of existing energy-saving technologies, they are unlikely to invest sufficiently in the cutting edge technologies necessary to achieve GHG reduction targets and prevent dangerous climate change. If firms are to invest in the long-term, high-risk radical technological innovations needed to mitigate climate change, they need regulatory clarity for years, in fact decades, ahead. In neither the US nor Australia are they getting this.

There is some cause for optimism though. Not only do large corporations in the US and corporate decision-makers in Australia generally accept the reality of climate change, they expect regulations and actually demand clearly defined rules. It is the uncertainty of anticipated rules and the potential for regulations to change frequently that is holding back investment and growth, not a fear of them. However, they do demand that any regulatory framework be clear, stable, and integrated across jurisdictions, both within and ideally across national boundaries.

Satisfying an environmental necessity is very different from satisfying a market demand and only government can frame the market processes to decarbonize the economy sufficiently to avoid dangerous climate change. As one Australian interviewee put it: 'I think it's really naive to expect that corporations, under the current structure that Western democratic society operates under, should or even can be the ones to take some sort of initiative in the absence of clear, strong, government regulation'. Either government has to directly invest in the development and diffusion of a range of technologies to reduce emissions by a defined amount, or it must so structure the institutions within which the market operates that sufficient number of firms anticipate profits from such investments. At present, as one Australian interviewee put it, 'government's not doing its job'. This would also seem to be the case in the US.

Appendix 1: Analysis of CDP reports

As shown in Table 6.7, we selected 47 reports from a range of eighteen industry groups, as defined by the CDP. In most cases we chose all the reports in an industry group that were publicly available. We did not attempt to access reports that firms wished to only make available to CDP investor members.

We subjectively assessed both risks and opportunities as 'low', 'medium', or 'high' based on the reported anticipated impact and expected probability of occurrence. Where assessed as 'medium-high' or 'low-medium', for simplicity we rounded them down to 'medium' and 'low' respectively. In scoring responses we also took into account comments made by the reporting entity elaborating on their indicated measure of the risk or opportunity. If several risks or opportunities are grouped together in response to a single CDP Question with different measures under Magnitude of Impact – for example, some might be indicated as 'low' and some as 'unknown' – we scored the response as 'low', 'medium' or 'high' as we judged most appropriate. This assessment was necessarily subjective because many companies reported several different risks or opportunities and assigned different impact scores to each. In most cases their assessment of risk or opportunity was entirely subjective so for our purposes, this approach is reasonable. We are not attempting a quantitative measure of probability of occurrence or a magnitude of an impact but are attempting to paint an impressionistic 'picture' of the firm's seriousness of purpose in pursuing climate innovation and the relative importance they assign to any risk or opportunity.

To assess average perceptions of risks or opportunities within industries or across all sample firms we scored Low as 1, Medium as 2, and High as 3 then averaged these scores. Where a firm gave no response to the question, no score was recorded and the question omitted from the average. However, if the firm noted a risk or opportunity but did not estimate it or if it considered the risk or opportunity unknown, we scored their answer as '0' and included it in the average. We take a measure of 'unknown' to mean that the firm may continue to assess the risk or opportunity but is unable to form a strategic or other business response. The Index of Severity in Table 6.4 is calculated in the same manner: an average of the subjective scores of reported measures of risk (which also are subjective). This index is intended to indicate the average *relative* risk perceived by CDP respondents.

The CDP reports reviewed were predominately for 2011. In several instances companies reported substantial reductions in energy intensity through 2010 or 2011 and claimed to have a continuing intensity target without stating one. It appears that 2011 was a 'watershed' year – it was listed as the end of a period of targeted emissions reductions – perhaps because of continuing uncertainty

Table 6.7 Company reports reviewed by CDP industry sector

Industry	Companies
Airlines	Delta; Southwest; United Airlines Continental
Automotive	Ford; GM
Construction and Engineering	Fluor
Defense and Aerospace	Honeywell; Rockwell Collins; United Technologies
Diversified Chemical	Dow Chemical; Du Pont; Eastman Chemical; PPG
Diversified Industrials	General Electric; 3M; Parker Hannifin; Textron
Electric Utilities	American Electric Power; Duke Energy; Exelon; Southern Company
Heavy Equipment	Deere; Cummins; Terex
Homebuilding	KB Home
Industrial Gases	Air Products; Praxair
Industrial Machinery	Pall Corporation; Eaton; Dover; Johnson Controls (from Auto)
Information Technology	Intel; Applied Materials
Manufacturing	Krueger
Metals and Mining	Freeport-McMoran; Alcoa; Cliffs Natural Resources
Oil and Gas Exploration & Prod'n	Anadarko; Apache; Noble Energy
Oil and Gas – Integrated	Exxon; Conoco
Oil and Gas – Services	Halliburton; Schlumberger
Steel	US Steel
Waste	Waste Management; Casella

about the economic conditions and regulatory trajectories in both the US and EU. Several firms reported targets that were expiring in 2011 or 2012 and were 'evaluating' their plans for the future. For example, PPG, a diversified chemical company, 'achieved its five-year absolute GHG emissions goal in 2011 and is currently evaluating potential new goals'. In such instances we scored their expiring target as if it would continue into the future.

Our data are limited to what firms have chosen to include in their public reports submitted to CDP. It is possible that for strategic reasons they have not reported or not fully described all the decarbonization projects that they have implemented. GE, for example, states that it has implemented 415 such projects with sixty-seven more in some early stage of implementation but they have not individually described each project. In addition, they claim more than $1.5 billion is spent in 'clean tech' R&D annually but do not describe the projects that may ensue from this investment.

Finally, firms self-select to report to CDP. Therefore, the CDP reports do not necessarily represent the views and strategic or operational responses of the majority of employers in the US.

Appendix 2: Australian interview questions

It is widely believed that technological innovation will be crucial to achieving mitigation of a dangerous accumulation of atmospheric GHG emissions. We call this 'climate innovation', and I am going to ask you some questions about your firm's initiatives in respect of it.

First, some introductory questions:
1. What would you say are the major technological innovations your firm is taking that will reduce its greenhouse gas emissions? Why has your firm taken these initiatives? How would you characterize these: incremental or radical?
2. To what extent would you say these initiatives are driven by a desire to specifically mitigate your firm's impact on climate change as opposed to other reasons?

Thinking generally:
3. How do you believe climate innovation may best be encouraged: government regulations, consumer preferences (market forces), or internal company strategies? Please feel free to expand on your answer. Any difference depending the country you're operating in?

Thinking about the government specifically:
4. Do you believe government policy shapes the strategic direction taken by your firm with respect to climate innovation? If so how? Has this changed the strategic direction taken by your firm over the last ten years?
5. Would you say the approach of the Australian government on climate change is helpful? If yes: in what way? If not: why not?
6. What do you believe the role of the government should be in encouraging climate innovation? [PROMPT: standards, taxes, subsidies, penalties, rewards, information].

Thinking about market forces specifically:
7. What major market conditions, if any, have prompted your firm to undertake climate innovation?
8. What changes have you noticed in consumer attitudes or demand in the last ten years with respect to climate change? Have these changed your firm's strategic decisions significantly? How?
9. Which is most important in shaping your firms' long-term strategic planning on climate innovation: consumer attitudes or actual consumer demand?

Thinking about your firm specifically:
10. Are there any internal reasons that your firm has decided to undertake climate innovation. That is to say, what are the internal strategic drivers of such decisions?
11. What do you believe are the major challenges facing your firm in the next ten to twenty years with respect to climate change? How do you think your firm will respond to these challenges?

Finally:
12. Is there anything else you would like to add before we conclude the interview?

Notes

1 Some industries are more threatened by anticipated regulations than others, as exemplified by the Obama Administration's proposed regulation of GHG emissions from operating power plants, electric utilities are most exposed to regulatory costs. Details of the plan are available at http://www.whitehouse.gov/share/climate-action-plan. For the first time it imposes emissions regulations on existing plants as the single highest category of stationary GHG emitters. The regulations would primarily affect coal fired power plants.
2 Operating in multiple jurisdictions, however, also may bestow competitive advantages. Firms that are active in Europe have had to learn how to report under the EU Exchange Trading System (ETS) and monetize carbon credits. If, as most firms expect, the US will adopt a comparable system for pricing carbon within the next few years, those MNCs with experience of the EU ETS and the other regulatory systems in other countries will have gained valuable operational and marketing experience.
3 Firms normally set financial targets in terms of measures such as internal rate of return (IRR) as a benchmark for all internal investment projects to meet to justify funding. The target IRR is normally set at some multiple of prevailing interest rates. The lower the IRR that is demanded the longer the payback period (which means more risk) or the less profitable the project is projected to be. Acceptance of a lower IRR indicates that the firm has non-financial reasons (or reasons that cannot be quantified) to accept a higher risk or a lower return on its investment.
4 Scope 1 emissions are all direct emissions from facilities owned or controlled by the firm, scope 2 emissions are indirect emissions from the same facilities such as through consumption of electricity and other energy, and scope 3 emissions are those caused in the supply chain to the firm's facilities. See definitions at the GHG Protocol site http://www.ghgprotocol.org/calculation-tools/faq
5 Annual percentage reductions are approximate for several reasons, primarily that some firms have set multiple absolute or intensity emissions targets for different aspects of their business and have expressed targets as a percentage reduction over several years that may not equate to an integer percentage reduction annually.

6 The Obama Administration has proposed strict limits on GHG emissions from stationary sources that would make most coal-fired power plants inoperable without an effective system of carbon capture and storage. However, it is not technically difficult to convert many coal-fired stations to burn gas instead and may be economically viable.
7 As opposed to the idea of a price on GHG emissions which in theory was accepted and welcomed.
8 One interviewee spent considerable time discussing the stronger manner in which the US government funded and supported innovation, and then acted to protect its corporations' intellectual property, by comparison to Australia's.
9 For example, see Christopher R. Knittel (2011), 'Automobiles on Steroids: Product Attribute Trade-Offs and Technological Progress in the Automobile Sector', *American Economic Review*, 101(7), 3368–3369.
10 Between 2000 and 2011 Australia reduced its energy intensity by 1.7 percent a year, less than the US's 2.1 percent. However, from 2010 to 2011 energy intensity in Australia *increased* by 6.7 percent. The required decarbonization rate for Australia through 2050 is 5.3 percent, almost identical to the US rate.

7
US Labour Unions and Climate Change: Technological Innovations and Institutional Influences

Dimitris Stevis

Despite their decline over the last three decades US unions in the private sector remain one of the most significant and organized segments of US society. Without their support effective climate policy will be very difficult to reach and, even more so, to implement. Yet, there remain deep differences amongst unions on climate policy ranging from opposition to strong support. That variability is to be expected given the variable position of unions within the US economy as well as their own organizational and political characteristics. The two questions animating this chapter are: What kinds of technological climate innovations have US unions advanced and what strategies have they adopted in order to induce firms to adopt these innovations? Are any of their proposals profound enough both as innovations and in terms of protecting the climate or are they rear-guard efforts to protect a declining membership? Second, what institutional factors help explain the variability towards technological climate innovations and associated strategies evident amongst unions? In particular, are their choices largely determined by external institutional factors or is there evidence that, while these factors exert an influence, unions are purposeful actors that can choose from a range of strategies?

The first part of this chapter offers an analytical scheme for differentiating amongst the various types of technological innovations supported by unions and outlines the types of institutional factors that are likely to influence the choices of labor unions in contemporary America. As will become apparent there is a wide range of strategies largely intended to support, persuade or induce corporations to adopt technological innovations in a manner that is not detrimental to labor unions and workers. The second and third parts outline the major technological innovations adopted by US unions and comments on the institutional factors that may have influenced their technological choices and the associated strategies to move them forward. These technological choices and associated strategies are deeply grounded in US liberal capitalism in which collaborative relations at the level of the firm and the direct involvement of the state, in the form of industrial policy, are exceptional rather than integral. Because of its significance the third part focuses on the BlueGreen Alliance (BGA), a joint effort by unions and environmentalists to address climate change by seeking to influence firms and modify the broader institutional context in the direction of more regulated, if not coordinated capitalism. I close by summarizing the findings and commenting on the implications for the goals of this volume.

Analytical considerations

In Chapters 1 and 2, the editors of this volume make a compelling argument that technological innovation is at the heart of climate innovation. Technological innovation, in fact, is central to all union approaches, albeit with major differences in terms of the breadth and novelty of the technologies that are being emphasized.

Unions and climate innovation

Some unions focus on a single technology others on a single industry or sector and others on the whole economy. In terms of novelty some unions call for improvements of established technologies, such as nuclear power, while others promote emerging technologies, such as renewables.

The strategies that unions use to induce firms to adopt these technologies also vary. Some involve direct negotiations and collaboration with firms, leveraging existing positive relations or seeking to create new ones, particularly in sectors in which firms need all the support that they can get. Another set of strategies seeks to influence the federal or sub-federal state to adopt policies that will affect the calculus of firms or whole sectors as discussed in Chapter 4. A third set of strategies seek to reform US capitalism in a green and socially responsible direction.

Rather than developing a new terminology, I think that we can differentiate the resultant types of responses to climate change in terms of eco-efficiency and ecological modernization (Jänicke and Lindemann, 2010) (see Table 7.1). This is justified by the fact that all climate

Table 7.1 A typology of climate innovations

Type	Characteristics	Examples
Narrow Ecoefficiency	Modification of existing technologies; Scale limited to particular activities or places	Carbon storage; Greening individual buildings; Limited attention to social impacts
Broad Ecoefficiency	Some new technologies; Covers parts of whole sectors or activities but largely mitigative or corrective innovation	Green building without spatial planning; Limited attention to social impacts
Weaker Ecological Modernization	Technical innovation covers large swaths of political economy but allows externalities; Aims to manage or reform political economy	A green manufacturing strategy that does not pay full attention to process or the whole production network; Attention to social impacts may be limited to particular constituencies
Stronger Ecological Modernization	Technical innovation covers whole political economy and relevant production networks; Aims to reform or transform political economy	A green manufacturing strategy that pays attention to process and production networks as well as social impacts

innovations seek to address climate change through a combination of technological innovations – as differentiated from ecocentric proposals that challenge industrial society, value smaller economic and social units and support some separation of nature and humanity (Warner, 2010). At one end are narrow eco-efficiency proposals that are articulated around existing technologies, single plants or facilities, emphasize limited technical fixes, and are largely at the discretion of the unit's principals. Broader eco-efficiency proposals would involve a company's whole production network, or a particular set of activities across a sector, adopt more stringent eco-efficiency standards, and submit to some kind of third party monitoring. At the other end are technological innovations that cover the whole relevant political economy and fuse social and environmental priorities, including the imposition of costs depending on power and responsibility. Such an approach has been labelled strong ecological modernization (Christoff,

1996). Weaker ecological modernization also tends to have a broader scale and scope, although it may cover sectors and not economies, and is not as stringent in its expectations and allows for significant externalities such as ecomodernizing one country or sector at the expense of another country or sector.

Institutional factors possibly affecting union preferences

As Brian Obach (2004) has argued unions, like other organizations, variously interpret the institutional arrangements within which they operate and respond to these arrangements in many different ways (for a comparative view across many countries, see Hall and Soskice, 2001a; Rathzell and Uzzell, 2012). For example, contentious industrial relations may lead unions to adopt contentious strategies or, as the United Steelworkers (USW) has done, work hard to engage willing corporations. I organize the factors that can affect the climate innovation proposals of unions into three nested categories: macro-institutional developments in the organization of the political economy and the role of the state, meso-institutional factors such as industrial relations and strategic social alliances, and micro-institutional factors such as the organization of the unions themselves (see Table 7.2). These factors are not offered as specific explanations as much as dynamics within contemporary US liberal capitalism that may affect the choices of labor unions. In the empirical narrative I highlight the factors that seem to play an important role while in the conclusions I draw the factors or combinations of factors that seem to influence the choices of labor unions.

The changes that have taken place in the US economy since the 1970s can be conveniently lumped into those affecting the composition of the political economy and those affecting its politics. During the last forty years the service and IT sectors have grown precipitously while manufacturing – the historically more unionized sector – has declined as a share of the economy (Gertner, 2011). Outsourcing and offshoring – sometimes from one state to another within the USA – have changed the geography of production and the balance between capital and labor (Handwerker et al., 2011). As a result unions are much less able now than during the era from the 1950s to the 1970s to pressure firms and to *demand* that new elements – such as climate innovations – be included in collective agreements.

Another element of US capitalism is the ambivalence of management towards direct and commanding involvement by the state. Capital's opposition is not directed at the active role by the state, *per se*, but to public policies that constrain managerial prerogatives. As the editors

Table 7.2 Institutional factors influencing union preferences

Level of Institutional Factors	Dynamics of Institutional Factors	Impacts on Workers and Unions
Macro-institutional	• Relative decline of manufacturing along with outsourcing and offshoring (globalization) • Limited and indirect role of the state in steering capital (industrial policy) • Limited social regulatory role of state	• Decline of manufacturing unions which dominated US labor for much of the post WWII period • Capital largely regulated through indirect means • State does not step in to ensure industrial peace or compensate for management-labor asymmetries
Meso-institutional factors	• Contentious bipartite industrial relations • Regulatory competition • No labor or socialist party to bring together unions and other movements	• Collaborative relations are the exception and the result of relations between specific unions and companies • Downward harmonization of labor standards • Unions must seek allies the relations with which are often not very profound
Micro-institutional factors	• Many unions in same sector • Unions often include members from different sectors • US unions largely liberal with business unionists focusing on serving their members and social unions seeking to open up to broader society	• Difficulties in coordination • Divergent priorities within unions • Business unionists not likely to push firms beyond economic issues; social unionists do not have the power or support to do so

suggest, there is strong evidence that states are very active in liberal market economies (see Chapters 1 and 2). As Block and Keller (2011) have argued the US federal state – and increasingly the sub-federal state (see Chapter 4) – plays an active role in the promotion of innovation and industrial policy (see Chapter 3). These strategies have been more prominent in particular sectors at particular times. For example, the defense sector for all of the post-WWII era and aerospace for much of it. In other sectors the role of the state remains fragmented, episodic, and more indirect than direct. During much of the 1980s and the 1990s the idea of economic policy and, even more so, an industrial policy of 'picking winners and losers' was anathema in policy circles (Contractor, 2012; Gertner, 2011). In recent years, however, the federal state and many sub-federal states have sought to facilitate and channel economic activities, albeit through economic incentives and disincentives and through targets that allow firms some or a great deal of flexibility. In that sense US state policies differ from those in coordinated and developmental capitalisms in which the state plays a central and more direct role.

These macro-institutional changes have taken place along with increasing anti-unionism. The US has always been characterized by antagonistic industrial relations but since the 1970s the attitude of business has become even more hostile (Hogler, 2004; Katz and Colvin, 2011). This does not mean that USA unions are radical; they are not. It means that the mechanisms of industrial relations – especially the emphasis on solving disputes through the country's antagonistic legalism – promote conflict more than collaboration. Collaborative relations do exist in the USA, albeit fewer than in years past. Yet, such relations are not mandated or enabled by broader institutions, as is the case in Continental Europe, South America or Eastern Asia. Rather, they are the result of efforts by a few unions and managements. Stated differently, the two sides must be willing and able to reach out to each other.

As a result, there is a checkered pattern of collaborative relations between unions and firms. In some cases there is evidence of state influence, particularly when it involves national security, nuclear power, or infrastructure. There the federal government can demand certain minimum labor standards and practices because firms accept and depend on federal funds. In other cases these collaborative relations are remnants of past practices, as in the automotive sector and telecommunications. In still other cases it is the result of union strategies to befriend and support willing firms, as in the renewable energy industry.

This pattern is reinforced by the fact that industrial relations in the US is fragmented, even more so than is the case in other liberal capitalist countries such as the UK (Bamber et al., 2011). The absence of country-wide corporatist relations is complemented by the absence of sectoral corporatism. This fragmentation is strengthened by the operation of two competing industrial systems in the US private sector. In simple terms the more union-friendly system has been prominent in the industrial states of the Midwest, Northeast and the Pacific Coast while the anti-union, right-to-work system has spread from the Southeast to the rest of the country.

Unions have systematically sought to influence the practices of firms, particularly with respect to labor practices but also with respect to investment priorities and location, by exerting pressure on the federal state and, increasingly, sub-federal states. Unions threatened by offshoring have sought rules, often as part of trade agreements, to limit incentives for offshoring and increase incentives for reshoring. In response to regulatory competition amongst US states they have sought higher federal labor standards and easier ways to unionize. Most importantly, unions have sought to persuade federal and sub-federal states to stimulate investment in various sectors through direct spending, for example in infrastructure and nuclear power; increasing demand, as with weapons and aerospace; and providing incentives and disincentives more directly to particular sectors. As discussed later, these tactics are evident in the case of the BGA.

In addition to relations between unions and firms and unions and the state one should also pay attention to the relations between unions and society at large. Over the years unions have developed close relations with certain religious organizations and networks, occupational health and safety practitioners, and so on. In recent years, and very much related to the present research, they have also developed closer relations with environmentalists. In contrast to relations with capital and the state that are unavoidable, relations with environmentalists have been a choice, underscoring the impact of the internal factors discussed below. At the same time, in the absence of a political party, the relations between labor unions and other social forces have been less institutionalized than they have been in European countries and other liberal capitalist countries such as the UK, Australia, or Canada.

The internal politics of labor unions is also an institutional factor that affects the capacity of unions to influence the priorities of firms. First, there are normally several unions in the same sector and, quite often, the same company. That creates serious problems of coordination and puts even a willing firm in a difficult situation. While the macro-economic developments of the last thirty years have contributed to its organizational decentralization this has been an element of US unionism from the very beginning. Second, over the last twenty years there has been a great deal of union agglomeration (Moody, 2010). The Oil, Chemicals and Atomic Workers, for instance, became part of the Paper Workers and that union part of the USW. Woodworkers are now in the same union as air mechanics and graduate students in the same union as automobile workers. Therefore, the same union may sometimes take diverse and often competing views with respect to the environment. In a case that I have looked at closely, two locals of the same union in the same state were at odds over that state's effort to adopt renewables. One local representing workers that would install and connect renewables to the grid sought to include certification standards and promote renewables. The other local, working for a utility that depended on coal energy was resolutely against the state policy.

Finally, the political preferences of unions must be taken into account, as the decisions of a number of them to engage environmentalists demonstrates. It is common to differentiate unions into 'business unions' and 'social unions' (Hrynyshyn and Ross, 2010). Business unions see themselves as service organizations whose goal is to get and often manage economic benefits for their members. Social unions have a broader view of the political economy and adopt stances on a variety of issues beyond the immediate workplace demands. This does not mean that business unions do not have broader ideological preferences. Rather, their ideological preferences are articulated around a commitment to liberal capitalism. Social unions, on the other hand, may range from more to less radical.

There is a vibrant debate as to whether social unionists are more likely to adopt environmental priorities than business unionists. There is strong evidence that social unionists often do not adopt environmental priorities and, in fact, do the opposite (Hrynyshyn and Ross, 2010). However, the US record suggests that it is more likely that social unionists will adopt environmental priorities as has been the case with gender and race priorities. In fact, the leading unions in the BGA are more social-unionist than the leading unions outside of it.

Union proposals for technological climate innovation

In order to provide the broader context within which the proposals below fall I must note that some unions have not taken a positions while networks within and across unions call for even stronger measures. Several unions have declined to adopt explicit climate policies. Of great concern are unions in industrial sectors with significant climate impacts, such as aerospace, that have remained silent. For example, the International Association of Machinists, the major union in the aerospace industry, has not adopted an explicit climate policy, largely following the relative silence of the industry. There is no mention of the environment or climate change on its website and the author has not seen any of their representatives in the meetings that he has attended. This contrasts with its more activist approach during the 1970s and early 1980s in support of renewables. At the other end some unions and networks of unionists have proposed very strong and ecologically centered proposals (Sweeney, 2012b). It is fair to state, however, that the proposals examined below account for the vast majority of the US labor movement.

Narrow eco-efficiency? Adaptation of old technologies

A number of unions have taken explicitly sceptical positions, some based on existential concerns about the impacts of climate change policy on the whole sector (Brecher, 2012). Most electric energy in the US is produced from coal, an energy source that is directly challenged by any strong climate policy or climate innovation. Coal mining companies, as well as the communities and states in which they operate, rail companies that transport coal, and utilities that use coal to produce energy constitute a formidable alliance. Unions along this production chain are an integral part of this alliance and have expressed their discomfort with practically all climate policy bills, on the grounds of their employment implications. Their preferred organization was Unions for Jobs and the Environment, 'an organization of 14 unions with more than 3.2 million workers in electric power, transportation, coal mining, and other energy-related activities' (UJAE, 2012). This alliance is now inactive, most likely because it largely served to lobby against pending climate legislation rather than proposing and undertaking broader practical initiatives.

The coal industry, including the related labor unions, has proposed a number of innovations based on existing sources and uses of energy to ensure that it remains part of the energy mix of the future (Banig and Trisko, 2011). Most important are its proposals for clean coal and carbon sequestration. While most analysts and unionists would doubt that coal can be made adequately clean or that sequestration is a long-term solution some public resources have been spent on these technologies. Were these innovations to be proven usable they would certainly prolong the role of coal and have a major impact on future climate policy.[1]

The unions in the fossil fuel sector are not the only ones that are sceptical of strong climate policy. A number of unions have cast their lot with nuclear power, continuing their historically close relations with an industry that enjoys a great deal of federal support (Savage and

Soron, 2010). These institutional arrangements have been in place since the emergence of commercial nuclear power and are likely to continue into the future. Starting with the Energy Act of 2005 significant amounts of public money have gone to the nuclear sector.

Important policy makers, including the present and previous Secretaries of Energy, many scientists, and various union leaders consider nuclear energy not only beneficial for the climate but, also, an ideal case for eco-efficiency employing smaller and safer reactors that run for longer periods of time. So far, the key unions supporting nuclear power have not developed a comprehensive vision of a climate-friendly nuclear economy but have largely followed the initiative of firms and the federal government. The bulk of their concern centers around opportunities for construction, skilled employment, and the preservation of long-standing relations with firms and the state.

In environmental terms there remain serious concerns about nuclear power despite the views of some prominent environmentalists and policy makers. In addition to costs, risks and wastes, nuclear energy also promotes extraction. The lines being drawn with respect to uranium mining in Western Colorado, for instance, are strongly reminiscent of the debates that took place during the 1970s. It is also worth noting that most local support for nuclear power is in the Southeastern US, a region that is conservative across a broad swath of issues and whose political representatives are amongst the strongest opponents of climate policy.

Fossil fuels and nuclear energy do not simply raise questions regarding their immediate climate impacts. In fact, one could make the case that certain wind turbine and solar panel production processes can also be very harmful because they also require the extraction of minerals, the use of chemicals – especially in solar panel production – and the disposal of used equipment. The broader problem is that fossil fuels and nuclear power are fundamental obstacles to an efficiency/conservation/renewables centred energy economy which, on the whole, is more likely to be climate friendly.

Broader eco-efficiency

Reflecting historical institutional arrangements, multiple internal constituencies, and anxiety about the future it is not surprising that some of the unions that hold sceptical views have also promoted proactive climate innovations. These are not inconsequential but they do stand out when contrasted with less-climate friendly policies adopted by the same unions. The Laborers' union (LIUNA), as a member of the BGA, supported a strategy to improve building efficiency, one of the major consumers of electricity. Other unions, such as the International Brotherhood of Electrical Workers, have also been very active with respect to electricity. LIUNA and other Building and Construction Trades (BCT) unions are very much in support of rebuilding the country's infrastructure but their green proposals in that respect are not as distinguishable from regular calls for infrastructural investments as are their proposals regarding energy efficient buildings. Towards that end they have undertaken training programs with contractor associations and have sought public financing and rules to promote green building.

While these unions are supportive of green building they do not have a comprehensive view of an economy-wide spatial reorganization necessary to enhance overall efficiency and conservation. Moreover, because of the close relations that they have with large contractors they are often advocates for developers that promote sprawl, albeit of green buildings.[2] That does not mean that there are no committed environmentalists within BCT unions. In a research project regarding the New Energy Economy in Colorado, for instance, Stratis Giannakouros and I came across a number of people who have a broader environmental vision for their union and industry (see Chapter 4). Yet, there are significant forces against such an approach within unions.

That many of the sceptical unions realize that climate presents profound challenges is evident by their commissioning a report (in collaboration with the Environmental Defence Fund) on climate-friendly industries (Gereffi et al., 2008). The project examined a number of products that could help the USA become a leader in climate innovation while increasing employment opportunities. There is no evidence that the report has led to any coordinated initiatives but it does signify that some sceptics are more concerned by the transitional impacts of climate policy rather than climate policy itself. In many ways this report, and the rationale behind it, is very close to the directions of the BGA discussed below. The fact that some of these unions, which are also in manufacturing, have not joined the BGA underscores the significance of their political choices.

Ecological modernization? The BlueGreen Alliance

The BGA is one of the most important developments in the relations between unions and environmentalists because of the range of participants, the longevity of their relations, and because of the progress the organization has made with respect to forging general and specific priorities. In my view it proposes a set of economic, social and environmental innovations that collectively fall well within ecological modernization as it is a profoundly technological innovation-based approach, albeit one shaped by the institutional dynamics of US liberal capitalism. The innovations that the BGA proposes cover most of the economy, speak directly to climate change, put a premium on the role of the state, seek collaboration with capital and the state, and emphasize technological innovations that directly and indirectly protect the climate. The BGA also offers itself as an example of the efforts of unions and environmentalists to forge an environmental agenda that is sensitive to both employment and nature. At the national level a number of unions have proposed agendas centering around green employment and a 'just transition' (Block, 2011; Rathzell and Uzzell, 2012). At the global level UNEP has taken the lead, in collaboration with the ILO and the International Trade Union Confederation to push for a global green deal along the same lines (Block, 2011; Luke, 2009; UNEP, 2008). So far the BGA has not been the subject of academic analysis (Kojola, 2009). For all of the above reasons I believe it is worth closer attention.

Origins and organization

Climate change rose in prominence for labor unions on the road to the negotiations that produced the Kyoto Protocol (Cohen-Rosenthal et al., 1998). Such was the significance of the issue that unions and environmentalists established a Blue-Green working group to broaden and deepen the collaboration between them that had emerged as a result of their common views on trade that had emerged during the 1990s around NAFTA and the WTO. Demonstrating the closer relations the discussions continued even though the American Federation of Labor – Congress of Industrial Organizations (AFL-CIO), respecting the wishes of a number of unions, rejected the Kyoto Protocol.[3] One important result of this search for common ground was a 2002 report that sought to explore the challenges and opportunities of climate policy on the economy with a focus on labor (Barrett et al., 2002). This report was the first comprehensive one – and the first to be the immediate result of the labor-environmental alliance but it was not the first effort to reconcile environment and employment (Cohen-Rosenthal, 1997; Cohen-Rosenthal et al., 1998; Grossman and Daneker, 1979; Renner, 1991). Towards the end of that decade calls for a green New Deal also received currency (Luke, 2009).

The election of President George W. Bush, and his Administration's general hostility towards climate policy, as well as unions and environmentalists, prompted a number of state-level initiatives which drew upon the foundations set by the rapprochement of unions and environmentalists. The Apollo Alliance was formed in 2001 to bring together political leaders, business, environmentalists and unions. The alliance's main goal was the creation of a more energy efficient and self-sufficient economy as a step towards enhancing the country's manufacturing competitiveness. Since its inception the Apollo Alliance has initiated a number of state level alliances that have promoted reforms at the sub-federal level. In 2008, motivated by the unfolding economic crisis and the prospects of a Democratic Administration, it published its New Apollo Program (Apollo Alliance, 2008). The program was comprehensive, covering buildings and infrastructure, revitalization of USA manufacturing, technology, and training.

Parallel with these developments unions and environmentalists formally launched the BGA in 2006 (Foster, 2010). Like the Apollo Alliance, unions and environmentalists had also turned to the states during the early years of the Bush Administration and they have continued along those lines to the present. For a variety of reasons the USW and the Sierra Club were not joined by others until 2008 but soon thereafter the BGA grew to include a number of unions and environmental organizations.

In 2011 the Apollo Alliance joined the BGA and its proposals have become integral to those of the BGA, especially with respect to manufacturing. During that same year the BGA went through a major crisis over the Keystone Pipeline (Sweeney, 2012a). All of the environmental members as well as a majority of the unions in the alliance were opposed to it. As a result of internal conflicts, the alliance did not take a position for some time. It finally came out in support of Obama's decision to delay the building of the pipeline, leading four unions to leave the alliance. The fact that two of those have now returned, suggests that the organization has achieved a degree of resilience. At this point the BGA consists of four environmental organizations (Sierra Club, the National Resources Defense Council, the National Wildlife Federation, and the Union of Concerned Scientists) and ten unions, including the USW, the United Automobile Workers (UAW), and the Solar Energy Industries Association (SEIA). The combined membership of the BGA is well over ten million people. An important development was the joining of the UAW along with the union's commitment to higher CAFE standards.

The composition of the Board of Directors of the BGA reflects both the origins and the broader reach of the organization. In addition to representatives from the member unions and environmentalists it also includes leaders of the Apollo Alliance as well as a representative from Ceres, a leading organization in the promotion of responsible investment. The BGA also has a Corporate Advisory Council whose members include some major companies such as Alcoa, Arcelor-Mittal, International Paper, AT&T, UPS, and others.

The BGA participates in a number of practical initiatives in collaboration with other stakeholders. The Clean Energy Manufacturing Center is a collaboration of the Manufacturing Extension Partnership of the Department of Commerce, the Solar Foundation, the American Wind Energy Association, and the SEIA. Beginning in 2009, the BGA started organizing an annual Green Jobs Good Jobs Conference that attracts environmentalists, unionists, local, state and federal government officials and political leaders, business associations in the renewables sector, and specific corporations that have taken initiatives in renewables. The vast majority of the public officials and political leaders who attend and speak are Democrats. However, one presumes that the business representatives cover the political gamut. Finally, most of the reports and the proposals of the BGA are the product of working groups that routinely include representatives from business, especially renewables firms, local governments, and community organizations.

Climate innovations

The priority sectors of the BGA reflect its composition but their treatment goes beyond the limits of any single member as its policy papers, reports, and press releases tend to take a comprehensive view of these sectors while placing them within the broader economy. In general, the BGA welcomes technological innovation but is quite sensitive to its implications for employment.

Clean energy is central (BlueGreen Alliance, 2012; BlueGreen Alliance and American Council for Energy-Efficient Economy, 2012). On the production side the BGA promotes renewables, especially wind and solar, while on the demand side it advances a number of proposals that will reduce waste and consumption as well as drive more clean energy production. To that end it supports federal and sub-federal Renewable Energy Standards (RES) as well as a National Energy Efficiency Resource Standard. The development of a country wide smart grid is high on its priorities.

The BGA advocates for revamping and expanding the infrastructure of the country. With respect to transportation it emphasizes public means, while cars and other vehicles have to be built cleaner and meet higher standards (BlueGreen Alliance and American Council for Energy-Efficient Economy, 2012). The BGA is strongly supportive of the higher joint fuel economy and greenhouse gas pollution standards. Faster and accessible broadband that will connect the whole country and facilitate economic and social interaction while cutting down on travelling and pollutants is also a key priority (BlueGreen Alliance et al., no date).

The overall climate innovation strategy of the BGA is reflected in its manufacturing initiatives, which are central to its vision (Apollo Alliance, 2009; BlueGreen Alliance and Apollo Alliance, 2011a, b, 2012). More broadly, manufacturing is at the core of fundamental debates in the US (Contractor, 2012; Gertner, 2011; Platzer, 2012; Pollin et al., 2008; Stokes, 2012). The US has never had a comprehensive industrial policy. After WWII, there was a partial industrial policy moved by the Department of Defense which continued into the 1960s and 1970s with the space program. During the 1980s and 1990s it became a dirty word but deindustrialization along with the rise of climate change has given advocates of industrial policy the opportunity to call for a green new deal. On the way to the 2008 election the prospects of a somewhat more cohesive industrial policy centred around clean energy and manufacturing received more attention (Block, 2011; Luke, 2009; Pollin et al., 2008). The dynamic was interrupted by the 2007 financial crisis but there was enough momentum to result in significant amounts of money targeted at clean energy in the American Recovery and Reinvestment Act of 2009 (ARRA). On balance, however, this was not an industrial policy act, either in resources or scope.

While federal intervention is considered necessary, the BGA and its predecessors have also targeted sub-federal states. On the one hand this initially reflected the Bush administration's hostility to proposals emanating from the broad Democratic Party and its constituencies. On the other it reflects a realization that states can be the engines of growth and the sources of emulation. The major challenge that the BGA has to face is the hesitance of the various states to adopt a cohesive and explicit industrial policy and, in many cases, hostility to including workers in their plans. That explains the fact that the BGA does not operate in South Carolina and Texas, two states with high potential for manufacturing but, also, great hostility towards unions (Bureau of Labor Statistics, 2013). It is not the case, of course, that sub-federal states do not have economic development strategies on which they spend a great deal of money. Rather, these strategies are not as cohesive as an industrial policy would be.

States can play a central role in financing. As the Alliance states, 'tax incentives and loan guarantees can provide the clean sector with the tools that it needs.... This support is critical since financing poses a major obstacle to rebuilding our manufacturing base' (BlueGreen Alliance, 2013). Thus, the BGA proposes the formation of green banks at the federal and sub-federal levels. In addition to some certainty with respect to loans, the BGA also calls for a longer-term approach to tax incentives and credits that support the renewables industry and other clean industries.

The federal and sub-federal state levels can also play a very important role in promoting a comprehensive manufacturing plan that moves the country towards a clean economy and reasserts its technological leadership. Collaboration between Universities, research centers, national laboratories, and firms in public-private partnerships is key here. Related to such a resurgence will be programs for training the new labor force. States can also play a very important role in ensuring that workers have rights and enjoy good and safe working conditions.

The national *Make It In America* plan (Apollo Alliance, 2009) can be considered as the framework within which the BGA's state plans operate. As the title suggests the goal of the plan is to revitalize manufacturing in the USA through a focus on clean manufacturing. The national and the state plans address the same issues, albeit with sensitivity to the specifics of each level of governance. Direct and indirect financing are central, calling – amongst other matters – for a Clean Energy Bank. Some of the financing 'should be set aside for local development agencies and other community-based or manufacturing support organizations' (Apollo Alliance, 2009). The funding should be attached to strong labor standards as well as domestic content of the products. Content standards should be sensitive to those places most affected in those regions hit hardest by unemployment.

In order to promote national production, the *Make it in America* plan calls for more support to be given to the Commerce Department's Manufacturing Extension Partnership. The Plan recognizes that without appropriate training these goals cannot be accomplished and so calls for increased funding for the Green Jobs Act which is part of the 2007 Energy Bill. Finally, the Plan calls for a Presidential Task Force on Clean Energy Manufacturing.

The Plan itself does not provide environmental goals and parameters within which it will be realized. Nor does it include processes as an integral element of green manufacturing. Were we to take is as a stand-alone program it would simply be a call for an industrial policy whose products are clean. However, the *New Apollo Program* is more explicit about environmental goals, even though it prioritizes climate innovation more than specific environmental goals. Even more broadly, the BGA is explicit about the significance of climate policy. Yet, the BGA's manufacturing approach is not one that starts from the environmental threat in order to find solutions but one that seeks to find solutions that can address both environmental and social threats.

Both unions and the environmentalists in the BGA have been critical of trade agreements that compromise environmental and labor regulations and allow carbon leakage.[4] Cognizant of the fact that there is a great deal of regulatory competition within the USA that can lead to carbon leakage, the BGA also supports strong sub-federal state-level rules, such as those in California, to be used both against other countries and US states. In general, carbon leakage offers itself as a common metric for unions and environmentalists, who have long been concerned about it. Now unions can also evaluate trade and investment in carbon terms. As a result the BGA calls for the federal state to negotiate international agreements and adopt domestic rules that level the global playing field and provide nascent industries with the time to become competitive. A closer look suggests that the BGA places more emphasis on manufacturing in the USA rather than manufacturing by USA companies. Foreign companies that establish their presence here and have a high domestic content are part of this equation. In that sense the BGA seeks to make the USA a destination for high-end production in green manufacturing and renewables.

All of the above efforts are directed at the federal or sub-federal states but are explicit about the key target, the behavior and practices of private capital. Stated differently, the BGA envisions an implicit kind of coordination that is mediated by the state through regulations, incentives, and (dis)incentives. The BGA does not envision a corporatist state, as is the case of CMEs, but it does envision an activist state within the parameters of liberal capitalism. Its approach, in fact, is not that different from that of UNEP and the ILO which promote a global social dialog for greener and decent employment (UNEP, 2008).

In addition to these state-mediated incentives for firms the BGA also places a great deal of emphasis on direct and collaborative relations with corporations in order to both bring them along and to exert pressure for a more activist state. In some cases these are corporations with which member unions have had long term relations: for example, the

Communications Workers of America (CWA) and AT&T, or USW and Alcoa. In other cases these are companies in the renewables sector and green chemistry. Here the unions have sought to support them while expecting better labor relations practices. The relationship between USW and the Spanish wind turbine manufacturer, Gamesa, is an example of such relations. Industry alliances are active participants in various collaborative arrangements with the BGA and participants in the annual Good Jobs Green Jobs conferences. Indicative of the BGA's efforts to engage firms is its Corporate Advisory Council which includes a number of companies that share some of the BGA's priorities and have good labor relations.

Central to the various proposals of the BGA is a common metric: how many good, green jobs will be created. I cannot go into detail about what is a good and what is a green job here (Pollack, 2012; Renner et al., 2008). In general, good jobs are jobs that provide workers with a decent living, good working conditions, occupational health and safety, and a good community to live in. While workers' rights, occupational health and safety, and public health standards will certainly affect the process of producing green products, the discussion of green processes has not been extensively developed. This gap suggests that some of the climate innovations promoted by the BGA may not be 'strong' in the sense that Jänicke and Lindemann (2010) use the term; that is, innovations that address environmental goals. With this in mind the BGA has argued that the various initiatives of the federal and sub-federal states, particularly ARRA, have resulted in a significant numbers of green jobs (Walsh et al., 2011) and that there is the potential for more of them over time. The BGA, in fact, recognizes that there are or can be green jobs across the economy such as public services, water, restoration and conservation, recycling, mitigation, health and safety, and so on and has developed policy statements on some of them (BlueGreen Alliance, no date; Tellus Institute and Sound Resource Management, no date). In general, some of these jobs will be in totally new sectors and activities, for example, wind turbines. Others will be made green by producing for these industries, for example, steel production for wind turbines. And, finally many others can be made green through appropriate changes, such as recycling (Tellus Institute and Sound Resource Management, no date).

An important challenge that the BGA is aware of is that many good, green jobs are in capital-intensive industries, such as wind turbine manufacturing, which employ a limited number of workers. For that reason the BGA pays a great deal of attention to bringing back or building the whole supply chain here in the USA (hence the content provisions) so as to provide as many jobs in capital intensive sectors as possible.

On balance, then, the BGA reflects a collaborative approach towards business and government, as well as the world of finance, provided that these share some of the priorities of the organization. The organization's broad aim, in turn, is to rejuvenate and redirect the whole US political economy along certain environmental and social priorities. In this it differs from those unions that focus on particular firms and even sectors. The BGA, in short, seeks to change the broader calculus of corporate decisions on climate innovation.

Findings and conclusions

The two central goals of this chapter have been to outline the range of technological innovations proposed by unions and the associated strategies to realize them and to explore the impacts of institutional factors on these technological choices as well as on the strategies of unions to shift firms towards these choices.

As anticipated, there is significant variability amongst unions on climate innovation. Some are silent or opposed, others have promoted innovations that many environmentalists would find oxymoronic, still others are in the front with respect to climate innovations in some sectors and far behind with respect to other sectors. A second finding is that there are unions which, in collaboration with environmentalists, propose significant climate innovations both in technological terms and in terms of modifying the broader US political economy. To directly answer the first question, therefore, labor unions do offer environmentally sound proposals of technological climate innovation.

The significant literature on transitions that has emerged in the Netherlands, in particular, underemphasizes the political (Meadowcroft, 2009) while not discussing any connections it may have with an earlier approach to socio-technical transitions which was particularly sensitive to the implications of technology for workers and society (Geels, 2004; Kern, 2011). As a whole the technological proposals and the strategies of labor unions pay close attention to the socio-technical implications of climate innovation in the older version of the term 'socio-technical' which emerged in an effort to understand the impacts of technological changes in the organization of labor in the UK (Cohen-Rosenthal, 1997; Trist, 1981).

Having said that, there is such wide variability in the technological preferences and strategies that calling them all 'socio-technical' is not intended to suggest that they are all equally valid but, rather, that even the narrowest of them contain serious social concerns – whether for a union of highly skilled workers or for a union that aims to reform the whole economy.

With these general understandings in front of us which factors or combinations of factors can help us better understand the lay of the land and the specific proposals of the BGA? Macro-institutional factors are certainly important. Broad changes in the political economy of energy clearly motivate the coal industry network, in general, and the United Mineworkers, in particular. Fears that nuclear energy will go the way of coal – largely due to concerns over costs and risks – does motivate the nuclear industry network, including unions that see opportunities for building nuclear plants and infrastructure and getting fairly high skill jobs. The decline of manufacturing is central to the considerations of manufacturing unions. The opportunity for employment in clean manufacturing has led the USW, for instance, to establish close relations with firms in that industry as it reemerges in the US.

Off-shoring and outsourcing are strong concerns of the manufacturing unions. This has led them to support reshoring and to pay attention to whole supply chains, rather than to stand-alone original equipment manufacturers, that depend on distant suppliers. In order to accomplish these goals they also want international trade and investment agreements that prevent carbon leakage and level the playing field.

Unions have consistently called for a more active role by the state, whether federal or sub-federal. They clearly recognize that some administrations are more likely to promote policies that pay attention to their interests at the federal level but they are not opposed to collaborating with anyone who shares their goals at the state level. It is worth remembering that the BGA is not working in Texas and South Carolina which are amongst the ten most promising states in terms of clean economy jobs. The most evident reason is the hostility of local governments and firms towards unions.

A hands-off state and the demonization of industrial policy has led the BGA and those who support industrial policy to call for a regulatory and enabling state that uses incentives and disincentives to support environmentally responsible growth. Here we see important differences amongst unions, with some calling for a proactive state to move us in the direction of climate friendly innovations while others are satisfied with the traditional role of the US state in funding public works. The decision of the USW to call for state support for clean manufacturing was not an inexorable choice. They could well have limited themselves to calls for support of steel for pipelines, infrastructure, nuclear plants and so on. In fact the USW has said that it will hesitantly support the XL Keystone Pipeline if the steel is made in the US – clearly there are important tensions within this forward looking union, as well. Despite all this, it has chosen to make clean manufacturing a national priority and to collaborate with environmentalists – against significant internal hesitation.

Relations between management and unions are also central here. Where the key firms are unionized, as is the case with telecommunications and automotives, unions seek to bring those firms on board. Equally important, however, has been a concerted effort to ally with the renewables industry at a time when both the industry and unions find themselves struggling against the old economy.

But not all firms and unions that have close relations with each other are willing to embark on the road to a clean economy. Unions in the construction and fossil fuel sectors have close relations with firms (or groups of firms) and the state, especially in the case of nuclear power. In both cases unions are followers, partly because they do not see a way out. It would be inaccurate to say that these unions are simply trapped. Rather, a number of them ally themselves with these industries because they have not considered alternatives or because they support the existing practices.

The sustained dialogue between unions and environmentalists – and the creation of the BGA – speak to the importance of relations with other societal forces. It does not seem possible to me that unions would have developed the range of proposals that they have (and this equally applies to environmentalists) were it not for the influence of environmentalists as well as the support they receive from them.[5] Despite internal tensions and external challenges the BGA is not a tactical initiative. Rather, it is the strategic outcome of a long process. As one looks closer at the BGA it becomes apparent that unions and environmentalists see it as a political alliance whose goal is to affect the flow of things in the US – as distinguished from a tactical alliance whose goal is to respond to a specific challenge.

The impacts of unions on climate innovation and the creation of a green economy are also affected by the organizational characteristics of unions themselves. The fragmentation of unions leads to unions competing rather than collaborating. That can often be the case in the building and construction sector where there are more than ten unions, but jurisdictional conflicts amongst unions are common in every sector. Bridging the gaps is necessary but not easy. Unions are also agglomerations of constituencies resulting in the same union adopting different positions on clean economy innovations.

Because of all of these factors US unions have to be more agential and proactive. The empirical record shows that there are a variety of options. While there is every reason why the USW or the UAW would respond to their declining fates there is no *a priori* reason why they would participate in an organization that proposes changes to the economy along the lines that the BGA suggests. For a number of decades, for instance, the UAW was opposed to higher CAFEs. Once its new leadership took over, the union switched and endorsed higher standards. A major impetus behind this change was government pressure – in the midst of saving GM and Chrysler. Another source was that new leadership and its strong commitment to collaborative industrial relations. In fact, all the unions in the BGA offer collaborative industrial relations in exchange for significant willingness to help firms flourish. Many analysts and unionists have criticized collaboration but there are varieties of it. In some cases unions simply follow firms. In this case unions and environmentalists are trying to be at the forefront of changing the US economy.

To summarize, then, the variable attitudes of labor unions towards climate innovation supports the view that unions are institutionally embedded social actors that make choices. There is strong evidence that they are variably constrained and enabled by their institutional context as there is evidence that they can often change it. That was the case with breaking down the racial and gender boundaries by some unions earlier than others. And that is also the case with climate innovation and the broader environment.

The findings of this study underscore the editors' view that firms play a central role in climate innovations, albeit nested within a network of institutions (as discussed in Chapters 1 and 2). In the case of a strongly liberal capitalist country like the US the state is less able and willing to directly lead or regulate. Proactive policies, therefore, require a combination of incentives and disincentives. Unions and environmentalists are part of that network of institutions but they also seek to push it in particular directions that are consistent with their goals. Their goals, in turn, are more social than economic in nature. This does not mean that they do not push for better or living wages but that these wages are intended for the reproduction of the workers, their families, and the communities, rather than to pay dividends to shareholders. Unions also want more stable employment, something that requires embedding firms more than they may wish. In that sense this case underscores the editors' claim that 'the quantity and quality of firms' innovations is also influenced by non-economic factors....' Stable labor relations certainly influence the calculus of companies in mature sectors. When AT&T or ArcelorMittal introduce innovations they have to take into account their relations with CWA and USW, respectively, particularly relevant provisions in their collective agreements. Companies in emerging sectors may be even more willing to engage unions, if they need their support in order to break into a hostile market.

For all practical purposes, the introduction of renewables in the US is not a market process, despite the fact that the US is a liberal market economy. It requires a supportive social alliance and willingness to take a long view – something that venture capital cannot do. As Block and Keller (2011) have argued, the federal state has played an important role while chapters in this book add sub-federal states and social alliances. The major difference between a liberal market economy, at least as in the US, and coordinated market economies is that the political dynamics towards that goal are more affected by conjunctural factors, whether the election of Reagan that led to the demise of renewables in the US or the responses to the financial crisis that have led to a new wave of opposition to an activist state. This chapter argues that despite this more recent backlash there remains a social alliance involving unions, environmentalists, firms and elements of the state that may still succeed in making some climate innovations a permanent, if limited, component of the US political economy.

Notes

1 The interesting thing to note here is that energy extraction, transportation, and use are capital-intensive industries that employ increasingly fewer people, many of which are not unionized. The decline in employment and unionization is not related to environmental policies as much as it is related to the increasing capital intensity of the sector and the hostility of managements to unions.
2 Most workers in the construction sector are not regular employees of firms. Rather they are members of unions. These unions negotiate agreements with contractors (most often associations of contractors) that stipulate the hiring of union labor. Going against a contractor or an association of contractors in one case can impact the longer-term relationship while placing the union contractor at a disadvantage against a non-union one.

3 The AFL-CIO has normally sought to take positions that do not go against the strongly felt priorities of important members. What is noteworthy in this case is that a number of affiliates, such as the USW, continued to support climate policy. The dynamics behind this divergence anticipate the varieties of climate policy discussed later on in this chapter.
4 Carbon leakage refers to US multinational corporations moving their operations to jurisdictions which have inferior climate emissions or environmental standards. It also refers to foreign corporations producing products in those jurisdictions for sale in the US.
5 The Sierra Club, for instance, has come out publicly in support of easier unionization rules as well as the immigration policies supported by unions.

Part III
Climate Innovation Across Borders

It is national governments that face the challenge of implementing policies to meet GHG emission reduction targets, and corporations in distinct national institutional contexts that have to invest in climate innovation. This is the case whether or not international negotiations succeed and set binding national greenhouse gas (GHG) reduction targets. But the negotiations have been international, the problem of climate change is global, and the corporations themselves are increasingly multinational in their operations, in the sense that they are said to be freed from the territorial 'shackles' of the governments of nations through their investment, production and employment decisions. Although the focus so far in this book has been on liberal capitalism specifically, and primarily on US liberal capitalism, what happens when we 'cross borders' in respect of climate innovation? This is the focus of the contributions in this part.

In Chapter 8, Jeffrey McGee focuses on the observation made in Chapter 1 that it is almost as if there has developed a 'Washington Consensus' for climate change mitigation, with a received liberal capitalist conventional wisdom that the market and market mechanisms will underpin the necessary technological innovation. What he demonstrates is that such a global liberal conception of the problem and the solutions to it are no accident. They are certainly not the result of a global acceptance that there is no alternative to viewing the challenge of climate change. Instead, the ideology of what he terms 'neoliberal climate governance' has been employed and promoted by the US not just because the US is the exemplar of the liberal economic version of capitalism, but because this suits the pursuit of US economic interests. Once again the economy dominates the environment. He analyzes and explains the US preference for minimalist government intervention both internationally and domestically with markets as coordinators of economic activity, and thus the predominant 'solution' for innovation for climate change mitigation. In this way he shows that what seems like a lack of political will on the part of the US, and therefore a lack of leadership, is instead a product of its material concerns and its ideological standpoint. In particular, he shows that during the years of the George W. Bush Presidency, the US deliberately gravitated towards the least interventionist end of the climate policy spectrum, and therefore has set the global agenda for market rather than climate innovation. If liberal capitalism requires more creative government intervention, as suggested in Part I, and there is corporate acceptance of this, as suggested in Part II, US government actions have actively moved international negotiations in the wrong direction, entrenching a liberal institutional basis for them that results in a conventional wisdom that sees 'market innovation effectively trump serious efforts at climate innovation'.

But there are other perspectives and other varieties in capitalist relations of production between states worldwide. Particularly with the rise of newly industrializing economies such as China and their new multinational corporations establishing themselves on the world stage, the predominance of US/Anglo-Saxon liberal capitalism is not taken for granted as it once was. These developing states and their corporations are emerging, and will continue to emerge, *together*, and the geopolitical transformation of economic and political power resulting from their emergence is important for the manner in which it challenges liberal economic perspectives in general and for addressing climate change in particular. In addition, whatever the dominance of the US in setting a global neoliberal agenda, China rivals the US as the major contributor to global GHG emissions, has the world's second largest economy, and therefore is equally crucial for achieving the climate innovation necessary for mitigating climate change. In Chapter 9, John Mikler and Hinrich Voss therefore examine corporate-state relations in the US and China as revealed through the reporting of their corporations. What they find is that regardless of the extent of their international operations, US and Chinese corporations offer quite different rationales for their environmental and social responsibility, reflecting the institutional environment of the states in which they are headquartered. The institutional 'imprinting' of their home states on their rationales for action affects their innovation strategies. Thus, while US corporations stress market forces and what the market dictates in addressing climate change (that is, the 'normal' market innovation framework outlined in Chapter 1), Chinese corporations stress the role of their home state and its development goals. As such, they find that the state is more likely to be able to produce climate innovation through 'its' corporations in China, *presuming this is a national priority*, than in the US where consumers and shareholder concerns remain paramount not just for the corporations, but for the US government.

Finally, in Chapter 10, Ian Bailey considers the interplay of regional and national institutions by focusing on the European Union (EU). He finds that the complexity and dynamic nature of institutions affecting climate innovation in the EU region confounds any simplistic rendering of them as more or less liberal. However, by focusing on France as the exemplar of a more state-guided *dirigiste* member state, versus the UK which is usually regarded as the most economically liberal member state, and doing so in the context of innovation for off-shore renewable energy, he highlights the evolving 'mélange' of institutional contexts that characterizes the EU. Building on the observation in Chapter 1 that climate innovation is much less predicated on market demand, and that off-shore renewable energy is not driven by this, he demonstrates the way in which the varying institutional basis for encouraging it in France versus the UK has produced different results. In addition to highlighting interplays between European, national, regional and private-sector institutions during the construction, contestation and interpretation of the EU's climate and energy policies, and the different methodologies used by member-state governments to decarbonize energy production, his chapter also draws attention to the complex and uncertain processes through which technological innovation strives to achieve commercialization and navigate other components of the regulatory landscape. In other words, his chapter highlights the complexity of the institutional evolution necessary to deliver effective climate innovation outlined in Chapter 2.

Taken together, the chapters in this part highlight the need to develop a more robust foundation for analyzing national capability for climate innovation that flows from both national and regional institutional foundations that underpin capitalism, as well as material realities. In every nation institutions need to be modified in order to stimulate climate innovation. In the concluding chapter we identify the specific institutions in the US, that embodies the most liberal expression of capitalism, that would have to be modified to generate the climate innovation needed to prevent dangerous climate change.

We also discuss which political entity has the institutional structure that is most likely to generate effective climate innovation and which, as a result, will be able to lead the world to a collective solution to climate change. Finally, we suggest opportunities for further research.

8
The Influence of US Neoliberalism on International Climate Change Policy

Jeffrey McGee

The United States is home to the world's largest economy. It is also the second largest national emitter of greenhouse gases (GHGs), contributing nearly 20 percent of yearly global emissions (US EPA, 2013). In per capita terms, US GHG emissions rank amongst the highest of the developed countries (Garnaut, 2008: 55). It has long been clear that effective international governance for reducing GHG emissions will necessarily require significant US participation. The US has been a leader in researching the science of climate change through sponsoring research within its high quality university and government research institutions and making contributions to the United Nations scientific body in climate science, the Intergovernmental Panel on Climate Change (IPCC). However, wider US engagement with the international climate change institutions has been significantly less positive. During the early 1990s the first Bush Administration was active in negotiations to form the first overarching international agreement on climate change, the 1992 United Nations Framework Convention on Climate Change (UNFCCC). During these negotiations the US successfully opposed initiatives such as the inclusion of a system of internationally negotiated, legally binding targets and timetables for countries to reduce their GHG emissions. Instead, the US advocated that each country pursue their own domestic goals, strategies and/or programs for reducing emissions (Bodansky, 2001: 29). Despite some support for binding targets and timetables during negotiations for the 1997 Kyoto Protocol, the US position on targets and timetables has largely been one of ongoing resistance.

Explanations of US resistance to binding targets and timetables have pointed to domestic constitutional and/or political restrictions and protection of national economic interests. For example, authors such as Skodvin and Andresen (2009: 263) point out that influential coal and oil industries are key sources of domestic opposition to US action to significantly reduce its GHG emissions (also see the analysis of Giannakourus and Stevis in Chapter 4). This domestic political resistance makes it difficult for the President to gather enough support in the US Senate for strong domestic legislation to reduce emissions. Harris (2009: 968) argues that one of the primary goals of US foreign policy on environmental issues, including climate change, is to protect and promote the US economy. On this view, strong US action on environmental issues will only occur if it coincides with distinct benefit to the US national economic interests. However, there has been significantly less analysis of the resultant US efforts to develop an alternative ideology to binding targets and timetables for emission reduction. This chapter argues that 'neoliberal climate governance' offers a useful lens to analyze the ideology developed by the US to pursue its interests in global climate governance, including resistance to binding emission reduction targets within the UNFCCC. Neoliberal climate governance is used here to refer to an approach to climate change policy that favors the *least intervention in the decision making of relevant stakeholders*. Neoliberal climate governance therefore promotes the development of institutions that provide *a framework for individualized private decision making in responding to climate change*, rather than institutions that select and pursue collective societal goals in this regard. I argue that this conception of neoliberal climate governance provides a useful lens on the ideology developed by the US to resist at an international level the binding targets and timetables model for emission reduction and related equity-based redistributive claims of developing countries. The ideology of neoliberal climate governance was used by the US to resist developing country redistributive claims and maintain pursuit of US economic interests.

The chapter proceeds as follows. Section 1 explains the current dominant ideas on neoliberalism in environmental governance and provides detail on the broad conception of neoliberal environmental governance. Section 2 provides a detailed history of US resistance to binding targets and timetables in the UN climate negotiations and the development of alternative domestic climate change policies by the George W. Bush Administration. Section 3 describes US development of several alternative institutions outside the UN climate negotiations over the last decade. Section 4 argues that viewing US engagement with global environmental governance through a broad conception of neoliberal environmental governance provides important insights into how ideology has been used to support US interests and shape alternative international climate change institutions outside the UN climate process. Section 5 concludes the analysis with observations on what lessons might be drawn for the future development of the international climate regime.

Neoliberal environmental governance

There is a burgeoning literature on neoliberal approaches to environmental governance (for example, Anderson and Leal, 2001; Mansfield, 2004; Heynen et al., 2007). Much of this literature focuses on domestic initiatives by government at *privatization* of common pool resources to create markets in tradable resource extraction or pollution rights (such as Dryzek, 2005: 121–137; Buscher, 2010). Authors such as Driesen (2009) and Heynen et al. (2006) have documented, particularly in the North American context, the marketization of environmental goods through government creation of pollution and resource extraction rights. At the international level, Bernstein (2002: 5) has documented the growing influence of liberal economic ideas upon international environmental governance, including climate governance, over the past four decades. Liverman (2009: 293) and Newell and Paterson (2010: 26–27) have detailed US advocacy for emissions trading and other 'flexibility' mechanisms during the Kyoto Protocol negotiations in the late 1990s. As international emissions trading, the clean development mechanism (CDM) and joint implementation (JI) under the Kyoto Protocol are institutional mechanisms that rely on the creation of a market for emission reduction credits, these authors have also described this policy as a 'neoliberal' approach to international climate governance. The common element in the dominant literature on neoliberal environmental governance is therefore close association between the creation and use of markets in environmental goods in responding to environmental problems.

However, Driesen (2010: 2) explains neoliberalism as a philosophy of least intervention which 'presumes that free markets are much better at allocating resources than governments' and 'implies that government should avoid regulation whenever possible'. It is only 'when governments must intervene, they should do so by creating new kinds of markets in new kinds of goods, such as newly available radio frequencies, electricity futures, or pollution credits'. In similarly describing neoliberalism as a philosophy of least intervention, Plant (2010: 6) indicates that a common thread across all neoliberal thought lies in support for what Oakeshott (2006: 484) and Hayek (1976: 15) refer to as *nomocratic*, as opposed to *telocratic* institutions. As Plant (2010: 6) explains, a telocratic society is one 'devoted to the pursuit of some overall end, goal or purpose'. The end of the society having been established, it is then a matter for government to select appropriate policy tools, which may include market-based policies and others on the policy spectrum, to pursue the particular end. In contrast, Plant (2010: 6) describes *nomocratic* institutions, endorsed by neoliberal thought, as follows:

> Nomocratic politics focuses on the idea of political institutions as providing a framework of general rules which facilitate the pursuit of private ends, however divergent such ends may be. It is not the function of political institutions to realize some common goal, good, or purpose and to galvanise society around the achievement of such purpose.... Neither Oakeshott nor neo-liberals are much given to using terms like 'the common good', but if there is meaning to such a term then for Oakeshott and the neo-liberals it means the framework of rules facilitating the achievement of private ends; it does not lie in some substantive, collectively endorsed moral goal or purpose in society.

The common thread amongst neoliberal thought is thus a commitment to ensuring the least intervention in the decision making of individual actors in a society. This occurs best through institutions that are designed simply to provide a framework for individualized, private decision making of individuals rather than institutions that facilitate the societal selection and pursuit of a collective goal. The outcome of nomocratic political institutions is simply the aggregate of individualized private decision making rather than any politically determined social end, goal or target.

Adopting this broad conception of neoliberalism, it is possible to broadly map commonly discussed domestic environmental policies for mitigating GHGs on a spectrum ranging from the most interventionist (that is, least neoliberal) to the least interventionist (that is, most neoliberal). Whilst there may be some disagreement as to the exact location of a particular policy in a given context, the following Figure 8.1 provides a broad map of the location of various climate mitigation policies based on the level of intervention in stakeholder decision making:

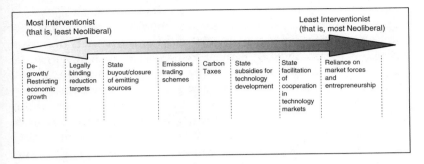

Figure 8.1 Climate change mitigation policy spectrum

Moving from the left of the spectrum, the most interventionist policies involve legal measures which place deliberate restrictions on economic activity so as to reduce the level of economic activity and GHG emissions emanating from a state. The recent global financial crisis of 2008–2009, while not a deliberate outcome of policy, clearly demonstrated that reducing economic growth is one way of reducing GHG emissions (Jha, 2010). The second most interventionist approach to reducing GHG emissions is through binding targets and timetables for emission reduction. This is a form of rationing of future emissions backed by legal sanction and requires strong state action to allocate the rationed emissions and police compliance with stakeholder emission reduction obligations. State buyout and closure of high emitting industrial sources of GHGs is also a highly interventionist policy measure. This involves using state funding to purchase, on a compulsory or voluntary basis, highly polluting industrial sources of greenhouse gas emissions (for example, highly polluting coal-fired power stations) that would otherwise be operated by private interests into the foreseeable future.

Emissions trading schemes also require a significant regulatory effort to establish an overall maximum level of emissions, create a system of tradable emission entitlements, allocate the initial emission entitlements and create detailed rules for calculating emissions and allowing trade in emission entitlements. The state effectively limits the activity of emitters by legally requiring them to hold sufficient tradable emission entitlements to justify their level of emissions in a given period. In implementing carbon taxes the state seeks to shift actor behavior towards lower emitting activities through placing a price on GHG emissions. Carbon taxes require a significant regulatory effort by the state through legislation to calculate emission outputs and the tax liabilities that individual emitters incur to the government. State subsidies for technology development involve the state providing incentives through schemes such as mandatory renewable energy targets or feed-in tariffs, to shift actor decisions on energy generation towards forms of generation with lower carbon emissions. State facilitation of cooperation in technology markets involves the state providing resources to overcome informational and coordination failures in private sector markets for cleaner technologies. This may involve the state establishing institutions to coordinate information gathering and sharing of best practices within industries and meetings to connect potential developers and users of technology. Reliance on market forces and entrepreneurship is the least interventionist approach to climate change policy and is located at the extreme right of the policy spectrum. This approach essentially involves allowing existing product and technology markets to respond to signals of consumer demand for reduction of climate change risk and adaptation activities.

As discussed above, the existing literature contains a dominant association between neoliberal climate governance and emissions trading schemes. However, as the climate change policy spectrum in Figure 8.1 demonstrates, there is ample space on the policy spectrum to the right of emissions trading schemes that is home to a number of policies that are significantly less interventionist than emissions trading schemes. The dominant association between markets in tradable emission rights and neoliberal environmental governance has therefore obscured the deeper neoliberal credentials of other less interventionist forms of climate governance that have played an important role, both domestically and internationally, over the last decade.

In the context of US domestic climate policy, few policymakers would advocate the policy options on the far left of the policy spectrum, such as deliberately restricting economic activity. However, policymakers located to the left end of the spectrum demonstrate a preference for the state limiting the range of economic actors' choices. Thus, government would increasingly insert itself into the decisions of economic actors and the state will increasingly regulate specific production methods/inputs and levels of emissions from sources (for example, regulating point source carbon emissions to make coal energy less competitive) through to the outright banning of certain economic choices (for example, no new coal fired energy plants). More centrally, it would subsidize certain activities that change cost-benefit calculations by economic actors in favor of lower emissions practices and technologies. Policymakers located to the right (that is, the neoliberal) end of the above policy spectrum prefer to let actors and markets work things out themselves. The right hand end of the climate policy spectrum is also consistent with what Mikler and Harrison describe as a 'market innovation' approach to climate change in Chapter 1. The market innovation approach provides that as demand for reducing GHGs evolves, producers are trusted to generate the necessary technological innovations to meet and profit from such demand. It trusts producers to develop new technologies that will incidentally provide a public benefit of a global reduction in GHG emissions.

Whilst the climate change policy spectrum in Figure 8.1 was derived from domestic policy approaches to reducing greenhouse emissions, the framework is equally useful for analyzing US engagement with international climate change institutions. As De Sombre (2011: 209) points out, in the case of the US, there are commonly important similarities between domestic environmental policy and the position taken in international environmental negotiations. The following sections detail US engagement with international climate governance over the last decade and argue that it displays a preference for least interventionist forms of governance located towards the right of the policy spectrum. US advocacy of these deeper forms of neoliberal climate governance was a key part of resistance to redistributive claims from developing countries within the UN climate negotiations.

US resistance to binding emission reduction targets

The UNFCCC and Kyoto Protocol

The 1992 UNFCCC[1] is the overarching international agreement that provides broad principles to guide the human response to climate change. The US joined the UNFCCC in 1994 after receiving Senate support for the treaty. The UNFCCC established a global goal of stabilizing GHG emissions at a level that will prevent dangerous climate change,[2] a general obligation on all countries to collect data on and report their GHG emissions[3] and burden-sharing principles to guide the future contributions and obligations of both developed and developing countries.[4] In a first experimentation with targets for emissions reduction, the developed countries listed in Annex 1 of the UNFCCC agreed to a *non-binding* commitment to reduce their GHG emissions to 1990 levels by the year 2000.[5] This non-binding commitment was agreed after an effective campaign by the US to prevent binding emissions reduction targets for developed countries being included in the final text of the UNFCCC (Bodansky, 2001: 29–33).

In 1995, a two-year period of negotiations under the UNFCCC was launched to set binding emission reduction targets for developed countries. Negotiations for these binding emission reductions targets were completed at the UNFCCC Third Conference of the Parties (COP3) meeting in Kyoto, Japan in 1997. The Kyoto Protocol (UNFCCC, 1997) provided a binding obligation for developed countries (that is, those listed in Annex B) to lead on reducing GHG emissions by making reductions in their emissions, as measured against a 1990 baseline, by a first target period of 2008–2012. The Protocol contained a target for the US, like most other developed countries, to effect an absolute reduction in GHG emissions of 7 percent below 1990 levels by the first target period of 2008–2012.[6] Under Kyoto, the developing world was exempted from binding emissions targets on the basis that implementation of the equity principle of common but differentiated responsibilities and capabilities, as agreed to in the UNFCCC, required that only the developed countries take on binding emissions reductions for the first target period. The Clinton Administration successfully argued at Kyoto for the inclusion of market-based flexibility mechanisms, namely, international emissions trading, JI and CDM, to allow developed countries to meet their emission targets at reduced cost (Depledge, 2005: 16–19). The US concern was to build flexibility into the Kyoto system of targets so that developed countries could take credit for reductions in emissions that occurred outside their own borders. The US argued this would be a significantly less expensive path to emission reduction for developed countries.

However, earlier in 1997, doubts had been raised over US participation in the Kyoto Protocol. The US Constitution provides the US Senate with a key role in the US entering international treaties. The Senate must provide a two third majority supporting vote before the President is able to ratify a treaty and bind the US in international law.[7] In July 1997 Democrat Senator Robert Byrd and Republican Senator Chuck Hagel passed a unanimous bi-partisan resolution through the US Senate ('Byrd-Hagel Resolution') indicating the following concern about the 7 percent emission reduction target that was being discussed as a possible outcome for the US at Kyoto:

> Whereas the Senate strongly believes that the proposals under negotiation, because of the disparity of treatment between Annex I Parties and Developing Countries and the level of required emission

reductions, could result in serious harm to the United States economy, including significant job loss, trade disadvantages, increased energy and consumer costs, or any combination thereof.[8]

The Byrd-Hagel Resolution clearly indicated the US Senate would not support implementation of the UNFCCC equity principle of common but differentiated responsibilities through differences in emissions reduction obligations of developed and key developing countries. Importantly, the Senate made it clear that the proposed absolute emissions reduction target of a 7 percent reduction in US emissions below 1990 levels would, in their view, cause harm to the US economy. In the clear absence of Senate support and growing domestic concern at the effect upon the US economy, the Clinton Administration failed in its efforts to have Kyoto ratified by the time it left office. Doubts over US participation in the Kyoto Protocol further escalated towards the end of the Clinton Administration. In late 2000, at the UNFCCC COP6 meeting at The Hague, the Clinton Administration abandoned negotiations on rules for implementing the flexibility mechanisms of Kyoto.

The George W. Bush Administration came to office in early 2001 and decided to withdraw from all further negotiations under the Kyoto Protocol (Depledge, 2005: 19). The reasons cited by the Bush Administration withdrawal were essentially those contained in the Byrd-Hagel Resolution, as expressed in a letter from President Bush to Senators Hagel, Helms, Craig and Roberts of 13 March 2001 (White House, 2001):

> As you know, I oppose the Kyoto Protocol because it exempts 80 percent of the world, including major population centers such as China and India, from compliance, and would cause serious harm to the U.S. economy. The Senate's vote, 95-0, shows that there is a clear consensus that the Kyoto Protocol is an unfair and ineffective means of addressing global climate change concerns.

This letter also contained the first strong indication that the US was prepared to become an advocate for an alternative international climate change institutions that would be more accommodating to its concerns (White House, 2001):

> Consistent with these concerns, we will continue to fully examine global climate change issues – including the science, technologies, market-based systems, and innovative options for addressing

concentrations of GHGs in the atmosphere. I am very optimistic that, with the proper focus and working with our friends and allies, we will be able to develop technologies, market incentives, and other creative ways to address global climate change.

President Bush's references to 'science', 'technology', 'market-based systems' and 'working with friends and allies' foreshadowed an alternative US vision for international climate change policy that would take shape in the second half of that decade. This alternative vision turned out to be a significantly less interventionist approach to international climate change that was first developed through the domestic climate change policy described below.

The 2001 Cabinet level review of climate change policy

In April 2001, following withdrawal from the Kyoto Protocol, President Bush ordered a Cabinet level review to examine 'science, technologies, current US efforts, and a wide range of innovative options for addressing concentrations of GHGs in the atmosphere' (US State Department, 2001a: 1). The instructions provided to the Cabinet level review were to develop 'innovative approaches' to climate change policy within principles that included the following: 'ensure continued economic growth and prosperity', 'pursue market based incentives and technological innovation' and 'based on global participation, including developing countries' (US State Department, 2001a: 1). The Cabinet level review released an Interim Report in June 2001 that contained a section dedicated to the US withdrawal from Kyoto. In this context, it is not surprising that the Cabinet level review made a central finding that Kyoto was 'fundamentally flawed' in that it 'fails to establish a long term goal based on science, poses serious and unnecessary risks to the U.S. and world economies, and is ineffective in addressing climate change because it excludes major parts of the world' (US State Department, 2001a: 13).

The absolute GHG emission reduction targets of Kyoto came under particular criticism in this report in that they applied only to industrialized countries and were claimed to be 'arrived at arbitrarily as a result of political negotiations and are not related to any scientific information or long term objective' (US State Department, 2001a: 13). More specifically, the 7 percent absolute emission target for the US is described as 'precipitous' in that it was based on a reduction below a benchmark level of 1990 emissions (US State Department, 2001a: 13). The Interim Report therefore suggested that any targets should be set against an 'emissions trajectory', essentially a business-as-usual case, thereby abandoning any reference to a benchmark year to gauge emission reductions and opening the way for a shift towards a softer type of target, known as 'greenhouse gas intensity targets'. These oblique references to an intensity target approach in the Interim Report gave a foretaste to its elevation in later US policy. The Interim Report highlights existing US Government programs on energy efficiency that have assisted in reducing the *carbon intensity* of the US economy by 15 percent from 1990–1999 (US State Department, 2001a: 3). The Interim Report only implicitly supports absolute emissions targets (even those based on a 1990 benchmark) when they are likely to be comfortably achieved (for instance national methane emissions and emissions from Federal Government buildings) (US State Department, 2001a: 4). While not clearly articulating GHG intensity as a new overall policy direction for climate change policy, the Interim Report provided the foundation for a later formal policy shift in that direction, as described below.

The 2002 Bush *Global Climate Change Policy Book*

In February 2002, the Bush Administration released a key environmental policy that explicitly articulated the 'new innovative' approach of the US as anticipated from the Cabinet level review. This document, the *Global Climate Change Policy Book,* carefully articulated the goals for US domestic greenhouse emission reduction and the new direction of US international efforts on climate change (White House, 2002). The opening words of the executive summary of the *Global Climate Change Policy Book* state that any climate policy acceptable to the US must involve no trade-off between reducing GHG emissions and ongoing domestic and international economic growth. The *Global Climate Change Policy Book* even suggested that economic growth is the solution to reducing GHG emissions (White House, 2002: 5):

> Sustained economic growth is essential for any long-term solution: Prosperity is what allows us to dedicate more resources to solving environmental problems. History shows that wealthier societies demand – and can afford – more environmental protection.

The argument behind this US statement is that a sustained period of economic growth is needed to make significant investment to decarbonize the US economy and allow an increase in consumption in developing countries in order to alleviate poverty. The *Global Climate*

Change Policy Book contrasts this 'pro-growth' US climate change policy with the Kyoto Protocol, which it characterized as 'penalizing economic growth'. It is implicit in this criticism of Kyoto that the US considered absolute emission reduction targets as a threat to its domestic economy. Further, the *Global Climate Change Policy Book* again made clear that the US would not support any implementation of the UNFCCC equity principle of common but differentiated responsibilities that required a redistribution of emissions and hence economic activity from developed to developing countries.

The centerpiece of US domestic action on climate change in the *Global Climate Policy Book* is a national target to reduce the GHG intensity of the US economy by 18 percent over the decade to 2012 (White House, 2002: 4), plus specifically linking GHG intensity targets to economic growth (White House, 2002: 4):

> A goal expressed in terms of declining greenhouse gas intensity, measuring greenhouse gas emissions relative to economic activity, quantifies our effort to reduce emissions through conservation, adoption of cleaner, more efficient, and emission-reducing technologies and sequestration. At the same time, an intensity goal accommodates economic growth.

The United States goal of an 18 percent reduction in GHG intensity over a decade was claimed to be 'ambitious but achievable' and equated to saving over 500 million tonnes of carbon emissions over the decade, against a business as usual case (White House, 2002: 5). The level of 'ambition' of this GHG intensity target was very questionable given that the natural rate (that is, without specific policy initiatives) of improvement in US GHG intensity over the period 2002–2012 was projected to be 14 percent. The *Global Climate Change Policy Book* goal of an 18 percent reduction in GHG intensity must therefore be viewed as only an extra 4 percent reduction in GHG intensity above the natural rate. Modelling by Van Vuuren et al. (2002) indicates that the target of an 18 percent reduction in greenhouse intensity would effectively allow the US economy to increase gross emissions to 32 percent *above* its 1990 level. Furthermore, the 'ambition' of the US intensity target must be assessed in the context of targets of industrialized countries under Kyoto that are mostly absolute reductions *below* 1990 national levels. The claims of 'leadership' of the United States on the issue of international climate change policy then start to sound very hollow. Clearly, any leadership by the Bush Administration over this period was more directed at protecting the national economy than reducing GHG emissions.

At the international level, the *Global Climate Change Policy Book* stated the US would pursue 'new and expanded international policies outside of Kyoto including building upon existing cooperative agreements on climate change scientific research and technology development with Japan and Italy' (White House, 2002: 3). Over the period 2001–2006, the US entered into bilateral and multilateral climate 'partnerships' with fifteen countries and regional organizations including India, China, Japan, Australia, Central America and the EU (US State Department, 2009). The US was also instrumental in the formation of a number of multilateral technology development partnerships aimed at climate change related issues. These multilateral technology development partnerships are the 'International Partnership for the Hydrogen Economy', 'Carbon Sequestration Leadership Forum', 'Methane to Markets Partnership' and the 'Generation IV International Forum' on nuclear power systems (US State Department, 2009). The unifying thread of these bilateral and multilateral partnerships was to increase the scientific understanding of climate systems, expedite cleaner energy technologies and develop better capacity for measuring and monitoring emissions. In contrast to the Kyoto Protocol, these US-inspired climate partnerships were non-binding and contained no international emission reduction targets or compliance mechanisms. The bilateral and multilateral partnerships thus continued a trend of the Bush Administration to favor voluntary, less-interventionist policies on climate change.

The next section discusses three key international climate change initiatives launched by the Bush Administration to provide less interventionist institutional alternatives to the targets and timetables model of the Kyoto Protocol.

US search for alternative international climate institutions

The Asia-Pacific Partnership

Over the period 2005–2008 the Bush Administration deepened the international strategy set out in the *Global Climate Change Policy Book*. A substantive move came in mid-2005 with the launch of the Asia-Pacific Partnership on Clean Development and Climate (APP). The partnership was announced at the 2005 Association of South East Asian Nations (ASEAN) Ministerial meeting with the six initial APP countries; China, India, Japan, South Korea, Australia and the US all present. In 2007 Canada was admitted as the seventh partnership country. South Korea was the most forthcoming with information in indicating the US was the lead country in initiating the APP (Government of Republic of South Korea, 2010):

> In 2005, the plan for organizing the Asia Pacific Partnership on Clean Development and Climate was proposed by the United States to the five Asian Pacific countries (Korea, USA, China, India, and Australia), which was increased to six countries by the participation of Japan in July 2005.

Similarly, the then Australian Foreign Minister, Mr Downer, received a question at the launch of the APP about which country proposed formation of the partnership. Downer (2005) responded:

> It was the Americans, well broadly...I mean the original initiative came from the Americans and they came and saw us...it was during the course of this year and we had some very good discussions with them in Sydney, and I mean it is up to them to talk more about it from their point of view...I think climate change is a problem and I don't think Kyoto is going to fix it, and so much political energy is invested in Kyoto for so little outcome and it has just sort of taken on a kind of ideology of its own.

A high level of US involvement in operation of the APP was also consistent with these South Korean and Australian claims of US initiation of the partnership. The US Department of State has housed the APP Administrative Support Group since its inception and the US has also been prominent in chairing or co-chairing four of the eight APP Task Forces (APP, 2010a). The US also contributed over one quarter of the public funding of the partnership during the period 2005–2008.

The Ministers at the APP launch claimed the partnership was designed as an 'innovative and a fresh new development for the environment, for energy, security and for economic development in the region' (Downer, 2005). They commented that the APP was based on 'policy integration' and 'technology cooperation' directed at climate change, pollution issues and economic development (Downer, 2005). Given the US and Australian presence in the APP, journalists at the launch asked whether the partnership was intended to be an alternative agreement to the Kyoto Protocol. Mr Downer was the first to state the official APP position that the partnership was intended to 'complement' the Kyoto Protocol rather than provide an alternative (Downer, 2005). Representatives of the partner countries have regularly repeated this official claim about the relationship between the APP and Kyoto Protocol over the past four years. The APP set no emission reduction targets for the partnership or for individual countries. Instead, each APP country was invited to set its own goals for emission reduction and encouraged to consider targets to reduce GHG intensity.

The APP Charter established a supreme governing body of the partnership known as the 'Policy and Implementation Committee' (PIC) comprised of representatives from the seven partner governments. The Charter also establishes eight sectoral (that is, industry-based) Task Forces comprised of representatives from the partner governments, public research bodies and the private sector (APP, 2007). It is the role of the APP Task Forces to formulate project plans for approval and funding allocation by the PIC. At the first APP Ministerial meeting in 2006 in Sydney the PIC approved over 100 projects for the eight Task Forces. In October 2007 a PIC meeting in New Delhi approved further Task Force projects, taking the total number approved to over 110, including eighteen 'flagship projects' (US Department of State, 2008: 7) designed to illustrate the potential and scale of APP projects (APP, 2009: 3).

By 2009, the total number of Task Force projects approved by the PIC was over 170 (APP, 2010b). The APP Task Forces met several times each year although the exact number and timing of these meetings was not made public. There were nine APP PIC meetings and three APP Ministerial meetings by the end of 2009 (APP, 2010c). The APP received only a total of $US200 million in public funding pledged by the seven partner governments. The APP expected the private sector to provide a significant amount of the funding for the implementation of APP Task Force projects. However, an analysis of the projects approved by the PIC indicates that the vast bulk of APP projects undertaken by the task forces related to gathering and disseminating information on industry best practices and facilitating dialogue amongst various industry and research stakeholders (McGee and Taplin, 2012: 316). The development of new technologies through joint public-private investment represented a relatively small proportion of the APP projects (McGee and Taplin, 2012: 316). In 2011 the APP was discontinued and existing projects that were not completed were transferred to other international technology cooperation institutions.

APEC Sydney Declaration

In 2007, Australia hosted the annual APEC Ministerial Meeting and Leaders Meeting in Sydney, Australia. At this meeting the Australian and US Governments attempted to reach an APEC wide position on a long-term, non-binding, global emissions reduction goal (Wilkinson, 2007a). However, key developing nations, particularly China, resisted attempts to agree on a global emission reduction goal outside of the UNFCCC process (Wilkinson, 2007b). The meeting produced a non-binding agreement titled the 'Sydney APEC Leaders Declaration on Climate Change, Energy Security and Clean Development' (APEC Sydney Declaration). Given China's reluctance to discuss global emissions goals outside the UNFCCC, the Sydney Declaration contained only a very weak stance by APEC nations to: 'work to achieve a common understanding on a long-term aspirational global emission reduction goal to pave the way for an effective post-2012 international arrangement' (APEC, 2007). The Sydney Declaration followed the APP in focusing policy on non-binding intensity based targets. It contained a non-binding target for a 25 percent reduction in energy intensity in the APEC economies by 2030, using a 2005 base year (APEC, 2007).

The APEC Sydney Declaration adopted an approach similar to the APP in attempting to shift the focus of international climate change governance towards a less interventionist approach based on voluntary commitments for research, information sharing and development of cleaner technologies. The Action Agenda attached to the APEC Sydney Declaration committed APEC to forming an 'Asia Pacific Network for Energy Technology' to strengthen cooperation between research bodies in the region in 'clean fossil energy' and renewable energy (APEC, 2007). The Action Agenda also committed the APEC countries to promoting clean coal technology and carbon capture through an Energy Working Group (APEC, 2007). The APEC Sydney Declaration is important in demonstrating how US sponsored climate agreements in the Asia-Pacific region were designed as a less interventionist alternative to the Kyoto binding emission targets and pursuing model. In doing this, the APEC Sydney Declaration continued the technology strategy, developed through the APP, of seeking to lessen informational and coordination failures in markets for cleaner technologies.

US major economies process

In 2007 President Bush launched a third US-sponsored climate change forum outside the UNFCCC in the form of the 'Major Emitters and

Energy Consumers' process (MEP) (White House, 2007a). Statements from the Bush Administration indicated the MEP was directed at 'both developed and developing economies that generate the majority of greenhouse gas emissions and consume the most energy' and was designed to address climate change 'in a way that enhances energy security and promotes economic growth' (White House, 2007a). The MEP involved US-sponsored meetings of fifteen of the world's largest economies and GHG polluters with a view to developing a long-term global goal to reduce emissions. Participating countries were expected to establish their own mid-term national targets and programs based on their national circumstances. The MEP also proposed that major emitting nations 'develop parallel national commitments to promote key clean energy technologies' with the US suggesting that international development banks provide low-cost financing options for clean energy technology transfer (White House, 2007a).

At the first MEP meeting in Washington DC in September 2007 the US described the proposed architecture for a post-2012 international climate agreement as based around a global aspirational long-term goal for reducing emissions, nationally determined policies to pursue emission reduction and energy security, sectoral based programs to reduce emissions, expansion of markets for clean energy technologies, action on deforestation and expanded financing for clean technology projects (US Department of State, 2007). The idea of a global aspirational goal for reducing emissions was a clear alternative to the binding emissions reduction targets of the Kyoto Protocol. In the MEP forum, the US clearly articulated a preference for a 'bottom-up' architecture for international climate change governance based on facilitating public-private partnerships for technology development and each country deciding their own emission reduction commitments (US Department of State, 2007). This bottom-up architecture required no binding targets or timetables and again provided an alternative, less interventionist and more deeply neoliberal design for international climate change governance. At the Washington meeting the US Treasury also proposed establishing an 'International Clean Technology Fund' (ICTF) supported by contributions from governments to help finance clean energy projects in developing countries (White House, 2007b). The ICTF was established under the administration of the World Bank and together with an associated Strategic Climate Fund has attracted pledges from developed countries of US$6.1 billion since 2008 (World Bank, 2008).

At the MEP meeting in Paris during April 2008 the US indicated its position on medium-term national emissions reduction commitments. The US proposal was to set its own goals to reduce the GHG intensity of its economy with a view to 'stop the growth' of US GHG emissions by 2025 (White House, 2008a: 2–3). However, this goal was highly conditional in that technology must advance sufficiently to allow the required emissions reductions to occur, again demonstrating the Bush Administration's focus on relying on technology development to allow emissions reductions. The final MEP meeting, held in Hokkaido in July 2008, produced the 'Declaration of Leaders Meeting on Energy Security and Climate Change' (MEP Leaders Declaration) (White House, 2008b). The MEP Leaders Declaration contains a 'shared vision' for a long-term cooperative global goal for emission reduction, but did not contain any attempt to quantify such reduction. The MEP Leaders Declaration indicates developed nations would implement economy wide mid-term goals and actions to achieve absolute emission reductions. The MEP Leaders Declaration also strongly emphasized the APP approach of sectoral-based technology cooperation and information exchange, demonstrating the inspiration it drew from the APP task forces and APEC in seeking to direct technology policy towards lessening informational and coordination failures in technology markets.

The following section makes the argument that the above US sponsored institutions (that is, APP, APEC Sydney Declaration and MEP) were essentially an effort by the Bush Administration to develop an alternative, less interventionist ideology for global climate governance that departed from the binding targets and timetables approach of the Kyoto Protocol and it's link with developing world redistributive claims through the principle of common but differentiated responsibilities.

Neoliberalism and the US approach to global climate governance

The above history demonstrates that during the Bush Administration's period the US was intent on developing an alternative institutional design for global climate governance. The US sought to develop this alternative institutional design in new forums outside the UNFCCC negotiations so as to avoid formal redistributive claims by developing countries through differentiation in targets and timetables for reducing greenhouse gas emissions. In these non-UN forums the Bush Administration sought to shift discussion towards ways that informational and coordination failures in technology markets might be relieved. Technological progress through spurring market activity became a preferred US substitute for political agreement under the UNFCCC to reduce or restrict US emissions.

The APP was emblematic of this US attempt at an alternative design for global climate governance. The APP followed a clear pattern in US international climate change policy during the Bush years of favoring voluntary commitments over the more traditional path of multilateral environmental cooperation through legally binding treaties. The APP was an important step in a wider shift from treaties to voluntary agreements in climate change governance that provided a reduction in the level of legalization and international political intervention on climate change. This process also represented an intended fragmentation of international climate governance away from the centralized international regulatory structure of the UN climate treaties and their model of binding targets and timetables. The APP was a particularly significant example of this fragmentation and reduced legalization as it was the first multilateral climate change agreement that suggested voluntary, nationally determined targets for reducing greenhouse emission intensity as an appropriate approach to goal setting on GHG mitigation. The APP therefore embodied a voluntary, fragmented and less interventionist approach to international climate change governance consistent with the deeper neoliberal approaches to climate governance shown in the policy spectrum in Figure 8.1.

Under the APP approach, the performance of technology markets was determinative of the level of ambition of national greenhouse intensity reductions and hence the ultimate global reduction in GHG emissions. The performance of technology markets thus replaced scientific recommendation and international political compromise as the key determinants of the level of climate change risk that would be allowed. The APP represented a nomocratic model of international climate change policy where market activity facilitated by state organized public-private partnerships essentially removed from international negotiations any global political compromise on key issues relating to global GHG emission mitigation. The APP offered the prospect of a technological approach to climate change policy that removed the necessity for internationally agreed binding emission reductions that carry potentially negative economic consequences. The APP thus embodied a model for responding to climate change in which the level of reduction in global GHG emissions would ultimately be determined by decision making of dispersed private actors in the operation of technology markets, rather than through collective global political compromise directed towards a common global end. The APP model of voluntary emission reductions and intensity targets was furthered by the US in the APEC Sydney Declaration and through the MEP.

In mid-2013 President Obama released a major statement on climate change to set an agenda on this issue for his second term office (White House, 2013). President Obama's 'Climate Action Plan' foreshadows a continuation of US efforts to selectively move international dialogue on climate change outside the UN climate process. The Climate Action Plan describes the UNFCCC as only one of several important international negotiations that the US will engage with on climate change (White House, 2013: 17). The Plan indicates the US will continue pursuing climate change negotiations through the Major Economies Forum (that is, an Obama Administration continuation of the MEP) and bilateral negotiations with key countries such as China and India. The Action Plan also indicates the Major Economies Forum will host an APP-inspired sectoral initiative to improve energy efficiency in the building industry (White House, 2013: 17). The US approach to international climate change policy flagged in the Obama Climate Plan therefore appears to be largely a continuation of the retreat from binding targets and timetables contained in the APP, APEC Sydney Declaration and MEP. The domestic mitigation measures contained in the Plan also contain no binding targets for national emission reduction. Instead, the Plan primarily focuses on using executive orders under existing legislation to: (i) reduce GHGs from existing power plants (ii) encourage energy efficiency and greater use of renewable energy (iii) provide standards for reduced fuel consumption in heavy duty vehicles (White House, 2013: 6). The modesty of these domestic mitigation plans confirms the Obama Administration's lack of confidence that the current US Congress would support more interventionist climate change legislation, such as a national cap and trade scheme.

The US strategy evident in the APP, APEC Sydney Declaration and MEP evidences a more nomocratic form of governance that requires significantly less international political agreement and intervention in existing market relations. The US approach also effectively passed decision making on the level of ambition of global climate change policy to the private decisions of actors in technology markets. The US strategy is thus located towards the right, neoliberal end of the spectrum in Figure 8.1. The US strategy is also consistent with the 'market innovation' approach to technological innovation discussed in Chapter 1 that is unlikely to deliver sufficient reduction in GHG emissions to avoid dangerous climate change.

Conclusion

This chapter makes a contribution to the literature on US participation in global climate governance and the wider literature on neoliberal environmental governance. It documents an important development in the US approach to global climate governance over the last decade developed by the George W. Bush Administration. Existing explanations fail to properly engage with the extent to which ideology mixes with material concerns to explain US climate change policy. This chapter therefore sought to advance understanding of US international climate change policy by showing that during the Bush years the US gravitated towards the least interventionist end of the climate policy spectrum. Earlier US advocacy of flexibility mechanisms, such as international emissions trading, gave way during this period to advocacy of a deeper form of neoliberal governance centered on voluntary international agreements directed towards improvements in GHG intensity and improving informational and coordination failures in technology markets. From the US perspective, these deeper forms of neoliberal climate governance had the advantage of sidelining developing world redistributive claims that might require the US to take a leading role in reducing emissions and funding adaptation. This US strategy sought to shape global climate governance by shifting discussions further to the nomocratic or neoliberal end of the climate policy spectrum where the outcome of mitigation efforts is determined by the private decision making of dispersed private actors. This US ideology would see market innovation effectively trump serious efforts at climate innovation. Market and economic concerns are placed before state intervention to effectively address climate change by reducing global emissions quickly enough to avoid dangerous climate change. US global climate policy during the Bush Administration period therefore represented a veneer of commitment to reducing GHG emissions, obscuring an institutional reality of a nomocratic, individualized, private regulation directed at market-led innovation. The recent announcement of the Obama Climate Action Plan confirms that we must wait yet further for the US to take the leadership role in climate innovation that the world desperately needs.

Notes

1. United Nations Framework Convention on Climate Change 1992, 1771, U.N.T.S, p. 107 (UNFCCC).
2. Article 2 of UNFCCC.
3. Article 4(1)(a) of UNFCCC.
4. Article 3(1) of UNFCCC.
5. Article 4(2)(a) of UNFCCC.
6. See Annex B of the Kyoto Protocol (UNFCCC, 1997).
7. Constitution of the United States of America. Article II, Section 2, Clause 2.
8. United States 105th Congress, *1st Session, Resolution 98*, http://www.gpo.gov/fdsys/pkg/BILLS-105sres98ats/pdf/BILLS-105sres98ats.pdf, date accessed 17 June 2013.

9
Varieties of Capitalism and US versus Chinese Corporations' Climate Change Strategies

John Mikler and Hinrich Voss

While climate change is global in its impact, its historical origins are in the world's major industrialized states, and they together with those that are industrializing are now its major contributors. This is why authors like Giddens (2011) have noted that casting the problem of climate change as fundamentally 'global' in nature abstracts from the geopolitical realities. The US and China in particular stand out for their contribution to the problem. They are both the world's largest economies and the two top greenhouse gas (GHG) emitting states, accounting for 40 percent of total GHG emissions in 2009 (World Bank, 2013a). Relatedly, there have been studies of the strategic responses by corporations to environmental issues. Of course, environmentalists and social groups often argue that profit-maximization oriented multinational corporations (MNCs) care little about the environmental consequences of their activities. But even if challenges in implementing environmental strategies persist, their environmental behavior has been presented more positively from strategic international business (Rugman and Verbeke, 1998a, b; Kolk and Pinkse, 2008) and integrative management-stakeholder relationship perspectives (Bansal and Roth, 2000). These studies, and others building on them, have enriched our understanding of the environmental responsiveness of corporations and highlighted the pressures on firms caused by market dynamics, environmental regulation, and societal expectations (for example, Bansal and Roth, 2000; Delmas and Toffel, 2008; Murillo-Luna et al., 2008; Darnall et al., 2010). However, such studies do not sufficiently stress the particular and differentiated national institutional contexts in which corporations are embedded when they respond to environmental challenges. This is a significant omission, because there is great institutional variance between countries' economic systems (Whitley, 1999; Hall and Soskice, 2001a; Redding, 2005; Jackson and Deeg, 2008), so that inevitably corporations and their environmental strategies evolve in conjunction with the institutional environments in which they are embedded (Peng, 2003; Mikler, 2009; Cantwell et al., 2010). And as noted in Chapter 1, corporate-state relations are key to understanding both climate and normal market innovation, and these must be understood in terms of the nature and effect of the nationally defined institutional contexts within which corporations are embedded.

Our intention in this chapter is to bring together institutional comparative political economy and international business perspectives on the motivators for climate innovation. We do so by focusing on United States (US) and Chinese MNCs and their home countries, primarily because they are from the largest economies with the greatest GHG emissions, but also because the United Nations Conference on Trade and Development (UNCTAD) Transnationality Index (TNI) – a simple composite average of foreign assets, sales and employment to total assets, sales and employment – shows them to be 'globalized' to different degrees. For example, the average TNI of US firms in the top 100 non-financial MNCs ranked by foreign assets was just fifty-eight in 2011 (UNCTAD, 2012a). Chinese corporations are, by and large, of a younger 'vintage' than their American counterparts and only just emerging on the world stage. There is only one mainland Chinese MNCs in the top 100 non-financial MNCs of 2011: CITIC Group with a TNI of twenty-three (UNCTAD, 2012a). As corporations internationalize their operations, their home bases remain important to them for the way in which the institutional basis of capitalist relations there informs their operations and corporate practices. But given the significantly larger TNI for American corporations and their more advanced age, a different interaction and influence from their home institutional environment can be expected.

Following on from investigating what motivates these corporations to take environmentally responsible action that potentially results in climate innovation, we ask if the managerial rationales offered for environmentally responsible action by US and Chinese MNCs is influenced by the more market-focused liberal versus state-guided environment that characterizes their home countries. Do the rationales they offer for such action differ, and if so how? Given the rationales that they offer, what are the pathways that are most likely to result in them undertaking climate innovation? These are the questions we intend to answer in this chapter.

The first section outlines the applicability of a comparative political economy approach to corporations' climate change strategies. The second section discusses the consequences of the different vintages of corporations from industrialized versus emerging market countries. The final section brings these two approaches together and applies them to the question of climate innovation in respect of MNCs headquartered in China versus the US by analyzing their corporate social responsibility and sustainability reporting. The analysis demonstrates that the national institutional contexts in which they are embedded means they stress different rationales for environmental responsibility, and from this we infer that they are likely to be on different corporate trajectories to achieve climate innovation. Management's motivation for climate innovation in US MNCs, which are based in and have emerged from a well-established market-focused liberal economy, are justified more on the basis of market signals and shareholder value. By comparison, Chinese MNCs, headquartered in an emerging and much more state-guided economy, stress a greater role for state and societal motivators.

A comparative political economy approach

The global operations of MNCs may be conceived of as differentiated networks operating across multiple, sometimes contradictory, institutional environments and covering a range of equity and non-equity relationships (Forsgren et al., 2005). This influences the way in which MNCs pursue their international strategies, in particular how their organizational and complexity-processing capabilities 'fit' with their host countries' institutional environments (Perlmutter, 1969; Bartlett and Ghoshal, 1989; Levy et al., 2007). But it is also the case that many studies have demonstrated that MNCs remain mainly organized around their home country or region and retain a strategically important headquarters (Rugman and Verbeke, 2004; Rugman and Oh, 2010; Buckley, 2011). Therefore, it is not just that individual firms have different ways of organizing their operations but that 'the basic institutional structures of MNCs may be influenced or even determined by the characteristics of states' (Pauly and Reich, 1997: 5). As Wade (1996: 85) put it, 'national boundaries demarcate the nationally specific systems of education, finance, corporate management, and government that generate social conventions, norms, and laws and thereby pervasively influence investment in technology and entrepreneurship'.

Therefore, MNCs' structures and innovation strategies are substantially influenced, if not completely determined, by the national institutional contexts of their operations.

In this vein, there is an extensive comparative political economy literature that stresses the manner in which there are variations in capitalist relations of production between countries as a result institutional complementarities that are largely socially determined (for example, Hollingsworth and Boyer, 1997), different national business systems as a result of states varying regulatory orientations (for example, Whitley, 1999), and states varying historical trajectories of development (for example, Dore et al., 1999). But in general, if institutions are 'a set of rules, formal or informal, that actors generally follow, whether for normative, cognitive, or material reasons' (Hall and Soskice, 2001b: 9), it follows that these establish different 'rules of the game' that may be thought of as 'the humanly devised constraints that structure political, economic and social interaction' (North, 1991: 97) and inform corporate strategies (for example, Heidenreich, 2012).

Perhaps the 'emblematic citation' (Crouch, 2005a: 442) in this regard is Hall and Soskice's (2001a) Varieties of Capitalism (VOC) approach which applies a two-categories framework in an attempt to distil the essence of the work of authors in the comparative capitalism tradition who studied the national institutional divergence of industrialized countries via multiple case studies, as well as the application of multiple categories that overlap but do not easily conform with one another (for example, Schonfield, 1965; Esping-Andersen, 1990; Crouch and Streeck, 1997; Hollingsworth and Boyer, 1997). It categorizes states as tending towards, or lying on a spectrum between, liberal market economies (LMEs) versus coordinated market economies (CMEs). Corporations based in LMEs prefer to coordinate their activities via market competition, are more 'arms-length' in their interactions with the state, and are more focused on consumer demand and providing shareholder value. By comparison, CME-based firms prefer more non-market cooperative relationships to coordinate their activities, including with the state and society. Of course, simplistically applying this typology to states abstracts too much from the reality of variance within the categories. The risk is 'a kind of rough, tough macho-theory that concentrates on the big picture and ignores detail' (Crouch, 2005: 452; see also Amable, 2003; Molina and Rhodes, 2007). Nevertheless, there is broad acceptance that the US is the archetypal LME, while Germany is the archetypal CME, and that emerging states such as China, and their corporations, face choices in the evolution of their VOC along at least two broad paths.

To some extent, the choice has already been made. While China's liberalization of its markets is an important feature of its emergence as a growing economic power, nevertheless its variety of capitalism is a state-guided one. For example, Beeson (2009; see also Halper, 2010) notes that China's 'state capitalism' means that the neoliberal orthodoxy underpinning the Washington Consensus is being replaced with a more pragmatic 'Beijing Consensus' that embraces the salience of state coordination. While it was once more readily accepted that ambiguous and emerging states would tend towards the LME category because 'the relationships of trust that are so central to the CME way of organizing an economy are hard to build and easy to destroy' (Goodin, 2003: 211; see also Streeck and Yamamura, 2001), China's rise clearly demonstrates that it is not just the economic dividends from the competitive self-interest underpinning liberal economic relations that are driving its economic success but its capacity for state coordination. We therefore agree with authors such as Fligstein and Zhang (2011), that China is an emerging market economy tending towards the CME category, or perhaps more accurately a state-guided *dirigste* version of capitalism, more in the vein of France (see also Tiberghien, 2007).[1] We recognize that sub-national differences at the provincial and city level exist which are caused by variances in the interpretation of central government institutional reforms and local institutional innovations so that 'we should not think of state-industry relations in China as national project organized in Beijing and implemented across the country' (Breslin, 2012: 41; see also Krug and Hendrischke, 2008). Nevertheless, China's state-guided capitalism means that management motivations for innovation generally, and in respect of climate change specifically, are likely to be less guided by market and shareholder imperatives, and more by the goals of the state and the national interest.

The vintage of MNCs

As within any organizations, MNCs have their set operational routines and organizational structures. They have an interest in preserving these because it helps organize and manage operations across a variety of cultures and institutional settings (Kilduff, 2005). Rather than adapting unopposed to a new context and morphing into local organizational and operational practices, routines that have supported growth and legitimacy in an MNC's home country tend to be retained (Suchman, 1995; Kogut, 2005). Because 'the national embeddedness of companies contributes to the reduction of uncertainties and to the solution of organizational coordination problems' (Heidenreich, 2012: 567), it follows that while possessing the capacity for adaptation to national contexts, as much as is possible they aim to deploy standard practices across their international operations, including environmental practices (Dowell et al., 2000). This produces a 'corporate organizational inertia' as whatever their potential for adaptation, corporations are best adapted to the institutional variant of the economic system they encounter at 'home', even as they internationalize their operations. A preference for operational forms that suit suggests the possibility of a path dependence in firms' host country adaptation strategies, as the relative ease of transfer of resources and capabilities established by an MNC in its headquarters is then a critical element in whether it can sustain competitiveness in a market initially unfamiliar to it (for example, see Kelly and Amburgey, 1991; Miller and Chen, 1994; Xia et al., 2009).

Partially following from this is the idea that an MNC is a bundle of organizational capabilities of which one is its ability to absorb and learn new knowledge in a particular context (Kogut and Zander, 1993; Kogut, 2005). The learning ability of an MNC, and therefore its ability to adapt and change, is crucial to its international performance. MNCs create path dependencies in terms of their international organizational structure as a reflection of their embeddedness in distinct national and regional contexts. Related to this is the development and deployment of resources and capabilities. Examples of these can be found in the current typologies of climate change strategies. Kolk and Pinkse (2008), among others, find that one successful corporate response to climate change is to embrace 'business-as-usual' and nurture existing capabilities. However, in contrast to incumbent MNCs, the current wave of new MNCs from emerging market economies such as China have not yet fully established what 'business-as-usual' internationally entails. Chinese companies that want to develop an international presence have made cross-border acquisitions in an effort to acquire technology and to learn the organizational structures and processes necessary to build a successful MNC. In other words, they have purchased innovation and innovative capacity rather than created it (UNCTAD, 2006).[2]

As such, these new MNCs are in the process of developing their international networks of subsidiaries and alliances and as they begin to invest across borders they have a wide range of organizational options for their international operations. In particular, they may mould their international presence around the key dimensions of absorbing and learning new technologies to upgrade their capabilities.

Indeed, it is widely argued that MNCs from emerging markets are lacking in the latest technological capabilities and international competitiveness and therefore invest overseas to better access the resources they require for upgrading in order to be more internationally competitive (for example, Dunning et al., 1998; Rui and Yip, 2008; Deng, 2009). If they combine different sets of knowledge accessed at home and overseas successfully, emerging market MNCs can potentially 'leapfrog' their industrialized county counterparts by developing and deploying capabilities that more immediately address and reconcile local demand structures. The case of the Chinese battery-turned electric car producer BYD illustrates this possibility. However, emerging market MNCs lack established channels of interaction and communication with stakeholders in foreign countries and, therefore, fixed routines of how to secure legitimacy, illustrated all too clearly in the CNOOC-Unocal case.[3]

The point to stress here is that Chinese MNCs' lagging or potential is a function of the process of institutional formation that characterizes the emergence of the Chinese capitalist state itself. As discussed above, it is difficult to categorize the China state based on existing typologies. As Peck and Zhang (2013: 2 and 3) note, 'China...has been a white space on the map of the VOC debate', because the VOC approach 'was developed exclusively in the context of "advanced capitalist nations"'. At its current stage of development, the Chinese state actually more closely resembles the developmental states of post-War East Asia (cf. Woo-Cummings, 1999; Johnson, 1995), or the 'strong' states of the eighteenth and nineteenth centuries that preceded the existing advanced, industrialized LMEs (for example, see Chang, 2002, 2003). Therefore, the role of the Chinese state in *leading* these corporations in the national economic interest is perhaps more salient than the role of the state as *coordinator* of economic activity (as per the CME category) or regulator of economic activity (as per the LME category). Indeed, Breslin (2012: 32) finds that there is 'a symbiotic relationship (at the very least) between state elites and many of the economic elites; they have effectively co-opted each other into an alliance that, for the time being, mutually reinforces each other's power and influence (not to mention personal fortunes)'.

The implication of this is that as Chinese MNCs are in the process of developing their capabilities, they must reconcile their operational interests with the demands of the Chinese government. As Wank (1998) and more recently Chen et al. (2009) note, relationships with party officials and political considerations override market and shareholder imperatives in a context in which the institutional environment is in a state of flux, with national development the overriding priority. As with previous developmental states, there is not a clear delineation between the state and market actors. Chinese corporations at this stage of the capitalist development of the Chinese state are therefore very much state-led and state-controlled. Therefore, their strategies, in respect of climate innovation as well as more generally, should be expected to be much more state-dependent than established MNCs from industrialized advanced states like the US.

MNCs' rationales for climate innovation

The distinct national institutional contexts in which the problem of climate change must be addressed has bearing for analyzing corporate environmental responsibility. Many authors have stressed that environmentally responsible behavior is a strategy for efficient and profitable business and delivering shareholder value (for example, Porter and van der Linde, 1995a, b; Moore and Miller, 1994). The ecological modernization literature also claims that there is a double dividend of ecological sustainability and profitability as a result of more enlightened business practices (for example, Mol and Sonnenfeld, 2000). However, Levy and Rothenberg (2002; see also Chapter 5) have observed that there is great heterogeneity in corporate strategies towards complex environmental issues. This suggests that while both material and institutional factors matter, the former is always predicated on the latter: 'market trends are themselves subject to institutional construction' (Levy and Rothenberg, 2002: 173). While it may be that factors such as regulatory compliance, shareholder returns, societal pressure, and economic efficiency are all generically important factors for how firms address their environmental responsibilities, our point is that this suggests that such factors are 'filtered' through the institutional lenses firms apply as a result of their institutional embeddedness in their home states, in this case the established corporations of the liberal US versus the emerging ones from state-guided China.

In order to examine the implications, we paired by industry the Fortune Global 500 Chinese MNCs listed on the Dow Jones Sustainability World Index (DJSWI) with US companies from the Fortune Global 500 that are also listed on the DJSWI. The industry sector classification follows Fortune (2012). This generated a sample of twenty-six MNCs, listed in Table 9.1. Firms are listed on the DJSWI when they are considered to be global sustainability leaders and belong

223

China	GHG Emissions metric tons CO_2e[b]	Global 500 rank[c]	Revenue US$ billion[c]	TNI/GSI[b]	US	GHG Emissions, metric tons CO_2e[b]	Global 500 rank[c]	Revenue US$ billion[c]	TNI/GSI[c]
Sinopec Group	NR	5	375	NA	Exxon Mobil	150,000,000	2	453	66
China National Petroleum	NR	6	352	3	Chevron	65,908,005	8	246	59
Industrial and Commercial Bank of China	NA	54	109	NA	Bank of America	1,709,890	46	115	34
China Construction Bank	NA	77	87	NA	Citigroup	1,075,929	60	103	74
China Mobile Communications	NA	81	88	NA	Sprint Nextel	2,027,545	328	34	NA
Agricultural Bank of China	NA	84	85	NA	JP Morgan Chase	1,323,591	51	111	34
Bank of China	NR	93	80	28[c]	Wells Fargo	1,601,048	80	88	NA
China State Construction Engineering	NR	100	76	31[a]	Fluor Corporation	NA	473	23	NA
China National Offshore Oil	NA	101	76	9	Conoco Phillips	70,200,000	9	237	40
China Life Insurance	NR	129	67	NA	Berkshire Hathaway	NR	24	144	37
Dongfeng Motor Group	NR	142	63	NA	General Motors	7,639,914	19	150	50
Ping An Insurance	NA	242	42	NA	American International Group	NR	109	72	35
Lenovo Group	91,593	370	30	43	Hewlett-Packard	2,000,826	31	127	60

Source: CDP (2012a, 2012b, 2012c); Fortune (2012); UNCTAD (2009, 2012a, 2012b, 2012c).
NR = no reported data
NA = not available
[a] 2007 [b] 2010 [c] 2011.

to the top 10 percent of the largest companies in the Dow Jones Global Total Stock Market Index. US and Chinese corporations account for 29 and 1 percent of the DJSWI respectively. Fortune ranks corporations globally by revenue, and this is also shown for each of the corporations in Table 9.1, along with their TNI and Geographical Spread Index (GSI). The GSI is very similar to the TNI but is reported for financial MNCs and focuses on the number and spread of foreign affiliates. The US firms have an average TNI/GSI of 49 percent, indicating that around half of their business is conducted outside their home country. The reported average TNI/GSI for Chinese firms is 23 percent but with very large variation within the group. It may also be noted that as well as being more internationalized, the US firms are older than their Chinese counterparts. The US corporations are on average 137 years old compared to thirty-nine years for the Chinese firms and the latter also have younger international operations. This is because Chinese outward FDI (and trade) was restricted until 1978 and only really came to life after the Chinese government issued the 'Go Global' policy in 2000 (Voss, 2011).

Also included in Table 9.1 are self-reported Scope 1 and 2 greenhouse gas emission data taken from the Carbon Disclosure Project (CDP). Scope 1 describes direct emissions from a corporation's owned or controlled sources (for example, production processes) and Scope 2 captures indirect emissions from the generation of purchased energy (Ranganathan et al., 2004). All the firms in the table provide data to the CDP but not all allow the data to be disclosed. It is, however, reasonable to assume that the emissions of the Chinese firms are likely to be somewhat higher than those of the American MNCs for corporations of similar size in the same industry, on the basis that US CO_2 emissions measured in kilograms per US$ of GDP (at purchasing power parity) are half of China's (World Bank, 2013b).

Environmental reporting published in English was then sought from these MNCs to ascertain how they thought their strategies should be 'best' presented to their readership. This was done explicitly because we recognize that such reporting is an exercise in public relations. It demonstrates how they defend and promote their actions and, therefore, from which their attitudes in respect of what they perceive as best practice may be inferred. Table 9.2 demonstrates that most corporations do not publish annual reports specifically on their environmental strategies, let alone their climate change strategies. Most include such reporting in their broader corporate social responsibility (CSR), sustainability or corporate citizenship reports. Six corporations (three from each of the US and China) do not produce any reports, or instead have

Table 9.2 MNCs' CSR and sustainability reports

China		US	
Sinopec Group	CSR Report 2011	Exxon Mobil	Corporate Citizenship Report 2011
China National Petroleum	CSR Report 2011	Chevron	Corporate Responsibility Report 2011
Industrial and Commercial Bank of China	CSR Report 2011	Bank of America	CSR Report 2011
China Construction Bank	CSR Report 2011	Citigroup	Global Citizenship Report 2011
China Mobile Communications	Sustainable Development Report 2011	Sprint Nextel	Corporate Responsibility Performance Summary 2011
Agricultural Bank of China	CSR Report 2011	JP Morgan Chase	Corporate of China Responsibility Report 2011–2012
Bank of China	CSR Report 2011	Wells Fargo	CSR Report 2011
China State Construction Engineering	No report	Fluor Corporation	Sustainability Report 2011
China National Offshore Oil	Social Responsibility Report 2010-2011	ConocoPhillips	No report
China Life Insurance	No report	Berkshire Hathaway	No Report
Dongfeng Motor Group	CSR Report 2011	General Motors	Sustainability Report 2012
Ping An Insurance	No report	American International Group	No report
Lenovo Group	Sustainability Report 2010-2011	Hewlett-Packard	Global Citizenship Report 2011

material on their web pages. The latter were not considered because considerable effort goes into publishing a written annual report, and therefore it presents what each corporation believes to be its *key* messages. While all the corporations examined have websites that contain a variety of information that evolve over time, these were not considered because a written report endures and presents, in one comprehensive document, the activities a firm believes are most important to communicate at a moment in time. Such a historical snapshot allows for a more robust comparative analysis on a like-to-like basis.

Sections of the reports considered were those where strategies and rationales for action were presented. These included the executive statements presenting the view of the Chief Executive Officer (CEO) and other board members that introduce the report; parts that could be considered overall strategic vision sections; and any sections that outlined strategic positions in respect of the environment, and climate innovation specifically. Codes were applied and analyzed using QSR NVivo 10.0 software for references in these reports to market forces, state regulation, society and internal company strategies, on the basis of the following rules:

- All coding was based on *rationales* for corporate action, not on the actions themselves. All coded statements answer the question of *why* action is being taken, rather than the simple fact that it is.
- Passages could be coded more than once. For example, a statement that it is necessary to respond to social concerns, and that in so doing market share will be increased, would be coded for both market forces and social concerns.
- Paragraphs were the maximum unit for coding. No coding was applied across paragraphs for the reason that each represents a new idea, or a new idea on the same subject.
- Sometimes the same code was applied more than once within a paragraph, if separated by a sentence/sentences that represented another idea. However, contiguous sentences expressing a rationale for action based on the same idea were not coded separately.

The coding results are shown in Tables 9.3 and 9.4 and are discussed below.

Market forces versus state regulation

A key difference was the extent to which US corporations stress reactive market drivers: maintaining profits and sales; reacting to consumers' demands; remaining competitive; providing shareholder value; and responding to market risks. They were twice as likely on average to mention these as rationales for action by comparison to the Chinese corporations. For four of them over 20 percent of the codes applied were for reacting to market demands. This was the case for *none* of the Chinese corporations. Proactive market drivers refer to rationales such as increasing profits and sales; increasing market share and market leadership; enhancing brand value; and grasping new business opportunities. It is notable that while a similar proportion of codes were applied to Chinese firms' reports for proactive market action, nevertheless on average the result was that market drivers accounted for 22 percent of the total codes applied to the Chinese corporations' reports, by comparison to 31 percent for those from the US.

The converse of this was the much greater extent to which Chinese corporations stressed complying with state regulations: 14 percent of codes versus 6 percent for the US corporations. In addition, Chinese corporations little stressed performing at a level beyond compliance such as through signing up to national or international voluntary agreements, and with one exception made no mention of providing input to the policy process. Therefore, while overall a similar proportion of codes were applied to the reports from both US and Chinese corporations for state regulation, the relative emphasis of the importance of this differed.

In addition to the relative intensity of codes applied, there were quite noticeable qualitative differences in the way in which the rationales were expressed between the US and Chinese corporations, but two in particular stand out. First, the references to market forces in the case of Chinese firms were much more obtuse. For example, the Bank of China's statement that 'we will continuously create value for our shareholders, customers and society through scientific development and accelerated transformation of development mode' is certainly a statement that the corporation sees a need to act in response to market demands. So is China National Oil's declaration that it 'has made contributions to the (sic) society by supplying energy to meet the (sic) demand'. Market forces are clear as a rationale for action, but they are hardly expressed as bluntly as General Motors' statement that 'being green helps us to sell vehicles', or the overriding focus on the company's bottom line implicit in Chevron's statement that 'we apply the same fundamental approach to our social investments that we apply to our capital investments'.

The implication is that the US corporations will embrace more market-driven climate change strategies than their Chinese counterparts. Being more focused on market forces and what the market dictates in addressing climate change, and reacting to this, they are likely to strategize on the basis of market signals more than the Chinese

Table 9.3 Chinese reports

	Market Drivers (%)		State Regulation (%)		Societal Drivers (%)		Internal Company Strategies (%)			No codes
	Reactive	Proactive	Compliance	Compliance + Policy Input	Reactive	Proactive	Internalization	History	Leader's Vision	
Agricultural Bank of China CSR Report 2011	3	0	20	0	3	27	30	17	0	30
Bank of China CSR Report 2010	18	18	18	0	9	18	14	5	0	22
China Construction Bank CSR Report 2011	9	22	17	0	9	22	17	4	0	23
China Mobile Sustainable Development Report 2011	17	6	0	0	11	28	39	0	0	18
China National Offshore Oil Social Responsibility Report 2010–2011	12	12	3	3	21	24	18	3	0	33
China National Petroleum CSR Report 2011	14	18	32	0	9	9	18	0	0	22
Industrial and Commercial Bank of China CSR Report 2011	14	6	17	3	6	33	22	0	0	36
Lenovo Sustainability Report 2010–2011	5	5	10	5	10	19	33	5	10	9
Sinopec CSR Report 2011	9	15	9	4	17	17	28	0	0	53
Donfeng Motor Group CSR Report 2011	0	13	13	0	47	13	7	7	0	15
Percentage of all codes applied to Chinese reports	**10**	**11**	**14**	**2**	**13**	**22**	**23**	**4**	**1**	**30**
Average of all codes applied to each company's reports	**10**	**12**	**14**	**2**	**14**	**21**	**23**	**4**	**1**	

Table 9.4 US reports

	Market Drivers (%)		State Regulation (%)			Societal Drivers (%)		Internal Company Strategies (%)			No codes
	Reactive	Proactive	Compliance	Compliance +	Policy Input	Reactive	Proactive	Internalization	History	Leader's Vision	
Bank-of-America CSR Report 2011	22	11	4	9	2	22	13	17	0	0	46
Chevron Corporate Responsibility Report 2011	42	11	5	5	0	21	0	11	0	5	19
Citigroup Global Citizenship Report 2011	8	30	0	3	8	43	3	0	5	0	37
Exxon Mobil Corporate Citizenship Report 2011	19	8	0	3	14	31	3	22	0	0	36
Fluor Sustainability Report 2011	9	4	9	5	4	21	9	18	14	9	57
General Motors Sustainability Report 2012	30	13	13	6	4	17	1	14	1	0	77
Hewlett Packard Global Citizenship Report 2011	19	21	8	5	4	21	5	15	3	0	78
JP Morgan Corporate Responsibility Report 2011–2012	26	0	0	5	0	42	5	21	0	0	19
Sprint Nextel Corporate Responsibility Performance Summary 2011	7	15	0	0	0	26	11	33	0	7	27
Wells Fargo CSR Report 2011	6	0	3	0	0	32	18	21	21	0	34
Percentage of all codes applied to US reports	19	12	6	5	4	25	7	17	5	2	46
Average of all codes applied to each company's reports	19	11	4	4	4	28	7	17	4	2	

corporations. By contrast, a relatively greater concern for state regulation, particularly complying with it, is the case for the Chinese corporations. This is further considered in the next section.

Societal and state forces

For societal drivers, differences are again noticeable. Although the percentage of codes applied is similar between the US and Chinese companies, for the former social responsibility was couched more in terms of reacting and being responsive to the needs of stakeholders directly related to the corporation's business (that is, customers, suppliers, employees, and the government) and responding to social concern, while for the latter there was much more emphasis on proactive responsibility to society, without social concern necessarily being expressed, and the need for the company to act in order to enhance trust, respect and generally high standing in a more general sense than brand value. In other words, there was less of a relationship between the corporations' explicitly *business* interests and those of society in a broader sense in the case of the Chinese corporations. This suggests that the Chinese corporations may invest in climate innovation to establish them as socially-acceptable in the absence of an explicitly business case, more so than their US counterparts.

In addition, there was a key qualitative difference in the tone of statements relating to state regulation. The US corporations often highlighted the need to comply with or exceed particular regulations. However, as well as citing compliance with state regulation more than their US counterparts, the Chinese corporations often discussed their plans not just in terms of complying with state regulations, but in light of the overall *goals* of the Chinese state. They referred to acting in the spirit of the country's twelfth five year plan and made statements along the lines of supporting 'the state's policy direction' (Industrial and Commercial Bank of China), and 'the call of the government' (Sinopec). The Agricultural Bank of China quite clearly states that it has 'earnestly performed the social responsibility as a state-owned commercial bank', including embracing the concept of sustainable development the objectives of which 'will be met in accordance with the cornerstone policy of "making progress steadily" put forward by the Central Committee'. There was a very clear impression of the Chinese corporations as substantively public, as much as private, actors.

This should not be too surprising because, with the exception of Lenovo, those in our sample are state-owned, and five of them are administered and monitored by the State-owned Assets Supervision and

Administration Commission of the Chinese State Council (SASAC).[4] Therefore, even when they cite responding to shareholders and addressing economic imperatives, they give the clear impression that they are doing so as much as a matter of national and social, rather than financial, interest.[5] It was interesting, too, to note that when it came to providing input to the policy process only China National Offshore Oil referred to 'interaction with international organizations'. There were no references to such input at the national level by the Chinese corporations. Therefore, by comparison to the US corporations that did this, for the Chinese corporations there was the impression of corporate responsibility flowing all in one direction: from the state to the company to society.

The implications are that the Chinese corporations show a greater preference than those from the US for climate change strategies in the absence of market signals and explicit regulatory demands, as well as for the wider societal benefit. They do this because they see themselves as servants of the state, as much as of their shareholders. By comparison, it would seem to follow that the internalization of social concerns such as climate change will require an impact to be made by either government legislation or regulations, or consumer preferences in markets for the US corporations. It logically follows that US MNCs' climate change strategies are much less likely to be proactively developed in the absence of market or regulatory imperatives both in their home countries and those in which they invest (also see data in Chapter 6).

Internal company strategies

Corporate strategies are developed based on internal beliefs and histories in addition to the exogenous factors related to the institutional context in which they are embedded. This should be particularly of significance for MNCs of an older vintage, such as those from the US, by comparison to those emerging relatively recently from China. But it was interesting to note that in stressing their histories of responsible business behavior and internal commitment to positive outcomes regardless of external drivers for this, US and Chinese corporations' statements were broadly similar. Even so, one point of difference was that three of the US corporations made explicit reference to this as flowing from the beliefs and vision of their leaders, whereas only one Chinese corporation did so. The fact that it is Lenovo, a company that was established by individual scientists outside the state enterprise system in the 1980s, and created by purchasing IBM's PC division (Ling, 2006), is perhaps notable in this case.[6]

Although in respect of internal company strategies there was little difference other than the lack of statements regarding leaders' personal commitments in the case of Chinese corporations by comparison to those from the US, it was also noticeable that overall on average twenty-six codes were applied in total to the reports of the former by comparison to thirty-nine on average in total for the reports from the latter. The relative intensity of their coding in respect of different rationales for action suggests that they have embedded their responsibility not just in different institutional contexts, but to differing degrees. The greater emphasis overall on providing reasons *why* the corporation has taken the action it has taken, as opposed to reporting the actual action undertaken, suggests an overall greater internalization of corporate responsibility for the US corporations.

The implication is that emerging market MNCs, such as those from China, are likely to relatively lack internally-generated climate change strategies because of underdeveloped internal motivators and relatively nascent external institutional pressures from international stakeholders. They have a preference for coordination, especially with their home state, but their capacity for this is relatively weaker than incumbent corporations from states that have moved beyond developmental objectives. As the institutional environment that underpins corporate governance and state-government relations in emerging economies such as China is in the process of being formed, so too are MNCs from them similarly in the process of developing a strong framework that could guide their legitimization on climate innovation and the development of the relevant capabilities. It might be hoped that compassion and individual concern – that is, a sense of social responsibility beyond simply brand enhancement – could fill this void but, as shown above, this does not appear to be the case for the Chinese corporations. The overpowering influence of the state and an orientation towards state development goals dominates their rationales for action, leaving relatively less scope for both internal drivers as well as market forces.

Conclusion

Institutions do not define outcomes. They do not define products or processes. They do not predict the relative success or failure of corporations' performance (for example, an MNC based in a more liberal economy may be as competitive or uncompetitive as one from a developing state-guided one), nor of their climate innovation. As such, variations in capitalist relations of production do not have explanatory power on their own. However, they do affect the strategic adjustment paths taken by MNCs, just as they affect the nature and implementation of the policy options of states. What we have demonstrated in this chapter is that how US and Chinese MNCs explain the rationales for their environmental and social responsibility is a reflection of the institutional environment of the states in which they are headquartered. This is 'imprinted' on their operations. It is a reflection of the national political economy of their home state, and relatedly their home states' stage of development and their vintage.

Corporations headquartered in the US stress market forces and what the market dictates in addressing climate change more so than those from China. The latter express greater concern for state regulation and state goals. Related to this, there is a relatively greater willingness expressed by Chinese corporations to cite social responsibility as a goal in and of itself by comparison to their US counterparts which instead see their social responsibility more in terms of market imperatives such as their corporate branding and financial interests. A greater embedding of the institutional basis for addressing their environmental and social responsibilities was suggested on the part of the US corporations given the greater number of codes applied to their reports. That is to say, incumbent US corporations are likely to have more established strategies that are perhaps a reflection of the deeper embedding of their home state's variety of capitalism, while those from an emerging China are in the process of developing theirs. US corporations' leaders appeared relatedly more personally committed to environmental and social responsibility than those of the Chinese corporations.

Of course, the extent to which this produces variations in corporate strategy is a matter of degree. It would be absurdly simplistic to make such pronouncements as a matter of dichotomous absolutes. However, our analysis does suggest different pathways to corporations undertaking climate innovation: market forces, market signals, and responding to regulatory and social imperatives in the light of these for US corporations, versus an enhanced role for the state and the national interest for Chinese corporations. Putting it simply, the state is more likely to be able to produce climate innovation through 'its' corporations in China than in the US where consumers and shareholders concerns are paramount. Our analysis provides avenues for future research, and the implications we raise need to undergo considerable empirical testing, particularly for the extent of the climate outcomes produced. Also for the extent to which they are more broadly generalizable beyond the MNCs and states considered. However, we would expect MNCs' climate change strategies to vary systematically, and therefore to find that the greatest potential for climate innovation lies with firms from more state-guided emerging market economies, in which the state has a greater role in promoting change and in which 'normal' market innovation is relatively less institutionally entrenched by comparison to those from established, industrialized liberal market economies.

Notes

1 Tiberghien identifies other East Asian states such as South Korea as sharing many institutional similarities with France for the manner in which the state and state elites direct, as well as coordinate, economic activity.
2 Noticeable acquisitions include Lenovo-IBM (USA), Geely-Volvo (Sweden), China National Offshore Oil Company (CNOOC)-Nexen (Canada), and Shanghai Automobile-MG Rover (UK).
3 CNOOC pursued its first major overseas acquisition with US-based Unocal Oil Company in 2005. Although Unocal's assets where mainly located in Asia and CNOOC offered a higher price than Chevron, it was outmanoeuvred by Chevron who heavily lobbied the US Congress which identified threats to national security.
4 It is interesting to note too that the US corporation which stressed compliance with state regulations most was the one that was bailed out by the US state and subsequently part state-owned in the aftermath of the global financial crisis: General Motors.
5 The Chinese state is also a major shareholder, holding significant portions of the tradeable and non-tradeable shares (Tian and Estrin, 2008; Li and Zhang, 2010).
6 Although which one of these two aspects may have triggered the inclusion is not clear.

10
Institutional Complexity in European Union Climate Innovation: European and National Experiences with Off-Shore Renewable Energy[1]

Ian Bailey

The influence of political and social institutions on business decision-making for the development and adoption of technological innovations has formed an increasingly important area of discussion among scholars interested in understanding the reasons for national differences in the direction and pace of innovation in response to societal challenges (Hall and Soskice, 2001a; Hancke, 2009; Nelson and Nelson, 2002). As Harrison and Mikler explain in Chapter 1, the influence of regional and national institutions on innovation processes is particularly evident in the case of developing technologies to mitigate climate change because of the manifold challenges involved in aligning scientific, policy and business priorities sufficiently to support the instigation, testing and commercialization of new greenhouse gas (GHG) reducing technologies. The greater difficulty, of course, lies in unravelling how such institutions and governance processes operate for different types of innovation, in different countries, and during different phases of innovation in order to arrive at a clearer view of the cause-and-effect relationships that determine whether or not innovations prosper or fall by the wayside. How satisfactory is it, for instance, to utilize the lens of liberal market and coordinated market economies (LMEs and CMEs) proposed by the Varieties of Capitalism (VOC) approach (Crouch, 2005b) compared with examining the political, social, economic and cultural institutions influencing climate innovation in individual countries and markets (Castree, 2006; Laurie, 2005)?

The dangers of glossing over the complexity and dynamic nature of institutions (the *variability* within varieties of capitalism) are especially acute when attempting to analyze the effects of institutions on climate innovation in the European Union (EU). The EU has been variously described as a system of multi-level governance and a federal polity in the making (Jordan and Adelle, 2012), but in basic terms it is a treaty-based organization whose member states have pooled some (but by no means all) of their decision-making powers. The treaties lay down the EU's general objectives, decision-making powers and policy procedures, while policymaking responsibilities are divided between a troika of supranational institutions, the Commission, Council and Parliament.[2] This governance system was designed in large part to prevent any single institution or member state dominating decision-making, but in practical terms means that, more often than not, EU policies are 'the aggregated and transformed standards of their original champions modified under the need to secure political accommodation from the powerful veto players...the resulting picture resembles less of a grand master plan...and more of a blend of many different elements – in short, a complicated "policy patchwork"' (Jordan and Adelle, 2012: 6).

The need to appreciate the institutional complexity of climate innovation in the EU is further underscored by the fact that the member states retain near-exclusive control of how agreed objectives are achieved. As such, climate innovation in the EU is affected by both supranational and national institutions. The two case studies examined in this chapter – the United Kingdom (UK) and France – in many ways typify the diversity of regulatory traditions that exist within the EU. France has developed a reputation for strongly *dirigiste* approaches to economic management, signified by strong indicative planning and state intervention to achieve strategic objectives. In contrast, the UK is often regarded as Europe's most liberal market economy (Schmidt, 2002). However, both countries' institutional and governance traditions have been deeply affected by EU membership. The UK has become more accepting of more regulatory forms of governance, while France moved towards more LME-like practices in some areas of economic policy, though in both cases, national institutions remain an important influence on policy development (Jordan et al., 2010). The result is an intertwined and evolving mélange of institutional contexts that reflects the EU's distinctive political arrangements as much as it does any individual market ideology.

The primary focus of this chapter is the effects of EU and national institutions on climate innovation in off-shore renewable energy and, in particular, the emergent wave and tidal energy sectors. However, care is needed when generalizing from analysis of one technology type.

In Chapter 1, Harrison and Mikler argue that climate innovations are often distinguishable from normal market innovations by a lack of *a priori* market demand for their emissions-reduction potential. As a consequence, different innovation processes are required that may include greater institutional and policy support in the form of legally-binding targets and/or financial incentives. These conditions certainly seem to apply to off-shore renewable energy, whose main challenges relate less to poor demand for its base product (energy) and more to: (i) a lack of proven record in providing abundant and reliable energy; (ii) the high levels of capital needed to develop, test and commercialize technologies and to connect facilities to grid systems (Vantoch-Wood et al., 2012); and (iii) the structural dominance of many national energy markets by large companies with heavy investments in fossil fuels (Musial and Butterfield, 2006; Mitchell, 2008). Additionally, opposition from local communities and other users of marine areas (for example, commercial fishing, shipping and recreational water users) on the grounds of visual intrusion, environmental impacts, or disruption to existing livelihoods may act as a deterrent to developers or investors where there is a risk of operating licences being refused (Bailey et al., 2011). One must also consider differences between the technologies that comprise off-shore renewables. Whereas off-shore wind power is achieving greater commercial competitiveness, most wave and tidal technologies remain in the experimental phase and are still generally viewed as high-risk investments (Haas et al., 2011). As such, patient finance and acceptance of uncertainty are at a premium and, more broadly, the institutional challenges facing off-shore renewables – and the range of actors involved in, or affected by, the sector – differ significantly from those facing low-energy appliances or motor vehicles. It is important that such differences are not overlooked in the urge to generalize.

Having noted these caveats, this chapter examines how institutions are shaping the development of off-shore renewable energy in the UK and France. It begins by examining the main EU institutional frameworks and processes affecting renewable energy policies in the member states and then explores the institutional contexts influencing the translation of EU targets in the UK and France. The main argument made is that the EU institutional context has provided an important catalyst for innovation in renewable energy in both countries but that national institutional preferences remain highly influential in the policies and measures developed to promote off-shore renewables. In particular, the chapter shows that both member states initially sought to 'fit' off-shore renewables into existing institutional and policy structures that were not well adapted to these new technologies, but have subsequently attempted to adapt these frameworks to meet the distinctive innovation needs of the sector. These emerging institutions nevertheless continue to bear the hallmarks of national governance traditions that struggle to meet the climate innovation needs of off-shore renewables. Emphasis is also placed on the wide range of actors whose involvement or cooperation is required to achieve the commercialization of off-shore renewables, in particular private (and public) investors and the various public and private actors involved in planning and consenting processes. The chapter concludes by reflecting on wider lessons gained on the role of institutions in stimulating corporate innovation in sectors such as off-shore renewable energy.

Off-shore energy in the European Union: Challenges and context

Over the past twenty-five years, the EU has sought to establish strong leadership on climate-change mitigation by promoting international action and through the development of internal targets and policies to reduce GHG emissions. The key elements of its internal policies on climate change include: (i) targets to reduce EU emissions by 20 percent below 1990 levels and to produce 20 percent of its energy from renewable sources by 2020; and (ii) the introduction of the EU emissions trading scheme and two renewable energy directives in 2001 and 2009 (Official Journal of the European Union, 2009). Although early policy attention in most member states focused predominantly on land-based renewable energy, recent years have seen growing interest at the EU and national levels in exploiting the region's potential for off-shore energy resources. Climate change and energy security remain the two major policy drivers; however, many national and subnational authorities have also identified off-shore renewable energy's potential to contribute to economic growth and employment in peripheral regions (Vantoch-Wood et al., 2012). As such, it is important to understand from the outset the multiple motivations for off-shore renewables that exist across different components of the EU's institutional framework and the potential effects of these on the evolution of institutional rules to promote innovation.

The first point to note about the EU's approach to promoting renewable energy is the use of directives as a principal legislative instrument. In accordance with the concept of subsidiarity – the general principle of European integration that specifies that the EU should only act when and to the extent to which objectives can be better reached by joint decision-making rather than by member states acting independently – directives require member states to meet agreed targets within specified timeframes but do not prescribe the methods used to achieve goals and, thus, seek to ensure that EU laws are implemented in ways that respect national circumstances and legal, political and sociocultural traditions (Jordan and Adelle, 2012). In this way, integration and sovereignty are balanced with the concept of the EU as a community of nations rather than a federal/federalizing polity. The 'tug' between collective action and national sovereignty is further demonstrated in the targets agreed under the 2009 renewable energy directive (see Table 10.1). Rather than adopting uniform targets, individual national targets were negotiated based on assessments of each country's renewable-energy potential, progress in developing capacity, and political discussions about the acceptability of national targets. Predictably, the directive's flexibility has led to national variations in the measures applied to promote renewable energy and the technology types and sizes prioritized (for example, commercial versus community or microgeneration) (Kitzing et al., 2012). The directive nevertheless stipulates that member states must submit action plans specifying how they will meet targets and 'indicative trajectories' for the period to 2020 (European Wind Energy Association, 2011).[3]

The 2009 energy directive also reveals the EU's role as a driving force for national action on climate change and, more directly, the use of agenda setting to galvanize climate innovation at the national level (Kellow and Zito, 2002). The EU's agenda setting role can be observed at several levels: its principled commitment to environmental issues and prioritization of climate change; its active role in international climate negotiations; and its internal policies. More specifically, experts have noted that the UK would not have developed ambitious plans for renewable energy without the 'push' of European legislation.[4] Table 10.1 shows that countries which sourced a lower proportion of energy from renewable sources in 2005 (for example, the UK and the Netherlands) accepted proportionately more demanding targets for 2020 than those that took stronger action by 2005 (for example, Denmark and Germany), though it also shows enduring variances between renewable energy 'leader' and 'laggard' states.

Agenda setting, of course, is not the only way in which political institutions can promote climate innovation. Despite a continued dearth of United States (US) federal-level climate-mitigation policies

Table 10.1 National renewable energy targets for 2020, by member state

	Share of Total Energy, 2005	Share of Total Energy, 2020	% Increase 2005–2020
Belgium	2.2	13	491
Bulgaria	9.4	16	70
Czech Republic	6.1	13	113
Denmark	17.0	30	76
Germany	5.8	18	210
Estonia	18.0	25	39
Ireland	3.1	16	416
Greece	6.9	18	161
Spain	8.7	20	130
France	10.3	23	123
Italy	5.2	17	227
Cyprus	2.9	13	348
Latvia	32.6	40	23
Lithuania	15.0	23	53
Luxembourg	0.9	11	1122
Hungary	4.3	13	202
Malta	0.0	10	n/a
Netherlands	2.4	14	483
Austria	23.3	34	46
Poland	7.2	15	108
Portugal	20.5	31	51
Romania	17.8	24	35
Slovenia	16.0	25	56
Slovak Republic	6.7	14	109
Finland	28.5	38	33
Sweden	39.8	49	23
United Kingdom	1.3	15	1054

Source: Official Journal of the European Union (2009: L 140/46).

(Harris, 2009), federal funding and tax credits for renewable energy and energy efficiency under the 2009 American Recovery and Reinvestment Act could be argued to be pursuing similar effects (Mendonça et al., 2009). The key difference between the two approaches lies in their respective emphases on policy *coordination* and policy *enablement* to promote innovation. EU renewable energy policy exhibits strong normative and procedural coordination, through the promotion of 'a European approach', bargaining over national targets, and supporting obligations (for example, interim targets and reporting), whereas the present US federal administration has relied on general normative steering (for example, high-level political speechmaking on climate and energy of the sort witnessed in Obama's second inaugural speech) intermixed with market enablement through the provision of financial incentives aimed at spurring technological innovation. As McGee demonstrates in Chapter 8, this represents a continuation of the policy settings and institutional preferences of the previous Bush Administration.

It is also worth noting that the EU's capacity to provide normative and procedural steering on environmental issues has had a long gestation dating back to the First Environmental Action Programme in 1973, when the EU held no environmental mandate and policymakers justified environmental initiatives as necessary to protect free trade (Bailey, 1999). However, growing experience of policy coordination on environmental issues before and since the incorporation of environmental policy into the Single European Act in 1986 has also contributed to a normalizing of the commitment to environmental protection, driven by a combination of: the actions of more environmentally-minded states along with the Commission and the European Parliament; the need to maintain a level playing-level for trade and economic policy; and the popularity of environmental issues with European citizens (Afionis and Bailey, 2012). However, Jordan et al. (2010: 9–12) argue that the EU's climate policy only really gathered momentum following its contests with the US over the Kyoto Protocol and the consequent need to fulfil the commitments made at Kyoto. Afionis and Bailey (2012) further note a growing belief that climate leadership could also help to achieve other goals, including spurring technological innovation, improving energy security and creating employment. Either way, the EU's commitment to coordinated action on climate change more closely reflects its political history and an ongoing blending of 'European' and national institutional traditions than it does any discrete European 'brand' of capitalism (Jordan et al., 2010). Moreover, many key decisions affecting climate innovation in practice remain under the control of the member states. Accordingly, the next section explores the significance of national institutions for climate innovation by examining attempts by the UK and France to promote off-shore renewable energy.

UK and French energy policy

Before looking in detail at UK and French policies for off-shore renewable energy, it is helpful to provide further background on the two countries' 'traditions' of economic management and energy policy[5] and the main policy issues affecting climate innovation in off-shore renewables in both countries. Turning first to national institutional traditions,

Mitchell (2008: 1) describes UK industrial policy as characterized by a paradigm in which the government provides: 'a regulatory framework that "steers" towards a defined general direction, then leaves it to markets to select the means to reach that end, with some regulatory limitations'. The emphasis, therefore, is on creating *enabling* conditions (Giddens, 2011) 'without old-style managerial intervention, and certainly without old-style public investment' (Mitchell, 2008: 21). However, although this policy 'style' that emerged as part of the shift towards neoliberal market economics initiated by Margaret Thatcher's Conservative government of the 1970s and 1980s has exerted a significant influence on UK energy policy, it has also become more fragmented following devolution, as the Scottish Government in particular has utilized its delegated powers to develop more interventionist regulatory and funding frameworks that are widely regarded as providing greater support for renewable energy than exist elsewhere in the UK (Scottish Executive Development Department, 2007).

French energy policy, in contrast, has been built on strong state intervention and a *dirigiste* style of state-influenced and state-sponsored entrepreneurship (Meritet, 2007). This was epitomized in government support for nuclear energy throughout the 1960s and 1970s, when it provided large quantities of investment capital and established operating companies for the sector (Prévot, 2007). The tradition of state interventionism to *ensure* desired objectives are achieved (in contrast with the UK's enabling approach) has also pervaded French policies for off-shore renewable energy (Kablan and Michalak, 2012). Although France has committed to more market-oriented energy policies as part of EU membership, significant challenges remain in reconciling the liberalization aims of EU energy policy with French conventions of state involvement in areas such as pricing and investment (Meritet, 2007).

Three main sets of institutional and policy issues can be identified for climate innovation for off-shore renewable energy. The first relates to national regulatory frameworks to support renewables, in particular the use of *financial instruments* to incentivize innovation and scaling up. The second centres on *assessment and planning policies* for evaluating and mitigating the environmental, social and economic impacts of off-shore energy projects. A key sub-component of the latter set of policies is the need for stakeholder consultation to ensure the views of relevant groups such as other users of marine areas and local communities are considered during decision-making. Identification of these processes also refocuses attention on the wide range of governmental, corporate and societal actors affecting innovation in off-shore renewable energy. These include: technology and project developers; finance companies considering investment in off-shore renewable energy; local authorities and development agencies seeking to attract investment and promote regions as centres of innovation; and local communities and other stakeholder groups whose interests are affected by off-shore renewables development (Arent et al., 2011). Having identified these contextual factors, the following sections compare the institutional processes through which the two governments have sought to promote climate innovation in off-shore renewable energy.

UK and French financial instruments to support off-shore renewable energy

Financial support policies for off-shore renewable energy can be split broadly into *incentive policies* and *direct public spending* to support infrastructure development and cost reduction in technology and supply-chain development. The fundamental purpose of both instrument types is to help manage the commercial risks faced by technology developers and potential investors until off-shore renewable energy can be driven to a greater degree by normal market demand imperatives. Among other things, this requires: financial rewards capable of offsetting the economic disadvantages of new technologies *vis-à-vis* more developed technologies (promoting patient finance); and clear long-term price signals that counteract the uncertainties of investment in prototype technologies. Both could also be said to give developers and investors greater decision-making autonomy by equipping them with stronger arguments to persuade shareholders and other stakeholders to invest in off-shore renewables projects.

Financial incentive policies

The primary UK financial support mechanism for renewable electricity since 2002 has been the Renewables Obligation (RO), a tradable certificate system that requires electricity suppliers in England and Wales[6] to submit specified numbers of Renewable Obligation Certificates (ROCs) equating to quantities of renewable energy generated, based on targets set by the electricity regulator. This target approximated to 3 percent of total electricity supply in 2002–2003 and has since risen annually by around 1 percent. The RO operates as a tradable certificate scheme, with ROCs being traded between electricity generators in an open market aimed at promoting target achievement at lowest cost. However, it also allows generators to pay a buy-out fine if they fail to acquire sufficient ROCs to meet their target, with the buy-out price effectively establishing a ceiling price for the ROCs market (eROC, 2011).

The main rationales given by the UK government for adopting a highly market-led approach were that competition and markets (that is, market enablement rather than enforcement) would provide least-cost solutions and avoid the government picking so-called 'winner' technologies that might not be the most climate or cost effective technologies (Mitchell, 2008). The main difficulty for more experimental and capital intensive sectors like wave and tidal was that standard ROC prices did not provide high enough incentives for investors to move away from more proven technologies like on-shore wind and solar energy. Furthermore, the variability of ROC prices in open markets undermined investor and business preferences for policy certainty. The UK government resisted calls to address these concerns until 2009, when it moved towards greater market steering by introducing a banding mechanism to increase the ratio of 'ROCs-to-electricity-generated' for selected less mature technologies. This resulted in wave and tidal stream receiving 2 ROCs MWh^{-1} and off-shore wind 1.5 ROCs MWh^{-1} (UK Government, 2009). The institutional context shifted further when Scotland raised support for wave and tidal-stream energy to 5 and 3 ROCs MWh^{-1} respectively; this was followed in 2010 by UK-wide increases for off-shore wind to 2 ROCs MWh^{-1} (UK Government, 2010). The government has since conducted a further review, which reduced support for off-shore wind installations built after 31 March 2014 from 2 to 1.5 ROCs MWh^{-1}. Additionally, support for wave and tidal stream technologies will increase to 5 ROCs MWh^{-1} for facilities up to 30MW capacity and 2 ROCs MWh^{-1} for larger developments (Department of Energy and Climate Change, 2011). Although each review was intended to balance maintaining investor confidence with promoting cost effectiveness as the commercial competitiveness of different technologies improved, the increasingly regular policy adjustments could equally be construed as fuelling business doubts about the government's long-term stance on off-shore renewables.

The investment context for renewable energy in the UK is undergoing further revision with the phased replacement by 2017 of the RO by Feed-in-Tariffs with Contracts for Difference (CfDs). Under this scheme, generators will receive electricity market prices plus a top-up to agreed 'strike prices', again banded by technology type. Where the market price for electricity exceeds this level, generators will be obliged to repay the surplus to ensure value and price stability for consumers.

The government envisages that revenue stabilization resulting from this approach will reduce investment risks and financing costs for innovative technologies, while still using competition to reduce costs and promote a level playing field between generating technologies (HM Government, 2012). However, the introduction of further – and more fundamental – policy reforms has added to existing uncertainty among developers and investors. Murray (2012) adds that the continued prioritization of the ideology of cost effectiveness may cause higher cost projects to struggle to compete despite their potential to deliver large volumes of low carbon energy in the long term. Additionally, the potential instability of strike prices caused by fluctuations in electricity spot prices, in conjunction with added policy complexity and uncertainty during the transition from the RO, may create further disincentives for investment.

In contrast to the UK's enabling approach, the main French incentive instrument for renewable energy places a stronger emphasis on ensuring renewable energy targets are met through the use of a feed-in tariff (FiT) system involving funding by contributions from electricity consumers and competitive tendering for target quantities of low-carbon electricity set by the public authorities and a multi-year programming of investments (Szarka, 2008). These are intended to provide certainty in the levels of non fossil fuel energy generated while providing stable income and cost streams for project developers and governments. Although there are similarities between the UK and French approaches to pricing mechanisms, in that both provide (or, in the case of UK CfDs, will provide) guaranteed prices and use competition to promote cost-effectiveness, important differences are also apparent. First, the French FiT scheme is less obviously tied to electricity market prices, so provides a greater guarantee of price stability. Equally, although the French FiT allows generators to decide what level of renewable energy to produce, price fluctuations are more tightly controlled by the state, while multi-year programming and tendering enables greater institutional steering of market processes. As such, the French system theoretically provides greater stability and confidence that renewable energy targets will be met, though controlling electricity costs depends on the government's ability to set appropriate repurchase prices and ensure genuinely competitive tendering (Meeus et al., 2012). Equally, both systems face challenges in managing competition between more and less mature technologies, and the temptation among investors to favour less expensive, short-term technologies over those with greater long-term carbon-saving potential. In this regard, finance ministries and their relationship with energy ministries becomes another important institutional factor determining price incentives for renewable energy.

Direct public spending

While the previous discussion provides some evidence of national institutional preferences influencing the use of financial incentives to promote renewable energy, the contrast is more palpable in respect of state investment. French governments have generally shown a greater willingness than their UK counterparts to commit state funds to research, technology, business development, and regional infrastructure for off-shore renewable energy. The French Great National Loan was launched in June 2009 to fund infrastructure projects that meet key societal needs. This has since been replaced by the *programme d'investissements d'avenir*, which includes a Renewable Energy Innovation Fund for off-shore renewable energy. Its first call in 2010 provided 142 million for projects over ten years. Management of funds is shared between the National Research Agency, ADEME (the Environment and Energy Management Agency) and OSEO, a state-funded body that oversees financing and loan guarantees for small-medium enterprises (Kablan and Michalak, 2012).

UK government investment in off-shore renewables remains modest in comparison, and is divided between several initiatives – the Energy Technology Institute, Technology Strategy Board, Research Councils UK and the Carbon Trust – each of which has its own remit and priorities. The now disbanded Regional Development Agencies (RDAs) also previously played a role in funding technology development for off-shore renewables, although their budgets were limited compared with those of central government (National Audit Office, 2010). Comparisons between the UK and French approach to state investment in off-shore renewables are complicated by differences in types of finance, timing and foci; however, the general view is that the UK has a complex and disjointed investment environment compared with the more centrally-driven funding landscape in France (Szarka, 2008). Differences are also evident in the scale of funding; approximately £620 million has been invested by UK government bodies compared with around £2.28 billion in France (Kablan and Michalak, 2012). Similar disparities in state investment can be seen at the regional level, where the *conseil régional de Finisterre* has invested heavily in port facilities and industrial platforms in Brest and Lorient. Although the UK's South West RDA also invested in infrastructure to support the Wave

Hub test centre in Cornwall (Bailey et al., 2011), RDAs have since been replaced by local enterprise partnerships that receive no funding from central government.

In terms of overall evaluations of the French and UK approaches to financial incentives for off-shore renewables, Haas et al. (2011) suggest that regulatory uncertainty and the risks created by price fluctuations mean that the UK's more liberal market has struggled to encourage investment or deliver economic advantages over more coordinated approaches. Similarly, the UK's Committee on Climate Change (2012) described the current government's funding for off-shore renewables as 'modest' for an industry with the potential to bring major benefits to the UK. In contrast, Kablan and Michalak (2012) argue that the French government's emphasis on direct funding and government oversight is critical in providing dedicated financial and human resources to validate and test technologies. The essential trade-off between coordinated- and liberal-market approaches nevertheless remains the level of importance placed on intervention to *ensure* desired outcomes are achieved versus facilitation to promote cost-effectiveness in the *pursuit* of desired outcomes.

Planning and stakeholder consultation

Off-shore renewable energy is somewhat atypical compared with most other climate innovations in that it requires large areas of physical space, either on land or in marine areas (Bailey et al., 2011). Planning thus forms a more prominent component of renewable-energy innovation, a fact that necessitates the creation of further institutional rules to manage relations between developers and the wide range of stakeholders potentially affected by off-shore energy, including regional and local authorities, local communities, commercial and recreational marine users, nature conservation bodies, and other agencies involved in promoting renewable energy or managing areas considered for off-shore energy.

As with strategic policy for renewable energy, the main legal frameworks governing assessment of the environmental and social effects of developments are developed at the EU level. These divide broadly into: environmental quality requirements (for example, the birds and habitats directives and marine strategy framework directive); and procedural requirements to undertake environmental impact assessment of projects with major environmental effects. Similar to the renewable energy directives and the EU's general approach to environmental governance, these frameworks prescribe broad requirements but delegate detailed management to national and sub-national authorities, leading to variable practices in the member states. For example, applications for off-shore energy projects in the UK are divided into two categories. Those under 100MW capacity are determined by the Marine Management Organisation, a non-departmental public body established under the Marine and Coastal Access Act 2009 to oversee the government's vision for clean, healthy, safe, productive and biologically diverse marine areas (Marine Management Organisation, 2012). Developments exceeding 100MW are classified as Nationally Significant Infrastructure Projects and are decided by the Secretary of State acting on advice provided by the Major Infrastructure Planning Unit, a subsidiary body of the UK Planning Inspectorate (Department for Communities and Local Government, 2012).

Consenting procedures for off-shore renewable energy have evolved in a more piecemeal fashion in France, with high levels of complexity and the involvement of a wide range of government agencies and regional/local authorities in planning processes. For instance, developers seeking permission for sites generating over 4.5 MW of electricity must obtain authorization from the Minister for Energy and the Marine Prefect to operate in the marine public domain,[7] a process that often involves consultation with local and national nautical committees, the Head of the Tax Services (to establish the financial conditions for concessions) and affected communes (Kablan and Michalak, 2012). In part, this provides a further indication of the imprint of historically-embedded French institutional preferences for state oversight that has established a relatively stable investment space but a rather unwieldy planning system. This procedure has undergone some rationalization following the adoption of the *Grenelle II* Act in 2010 and the publication of a new seaboard strategic document promoting integrated and joint management of activities in marine and coastal areas. This operates in conjunction with the Blue Energy Plan adopted in December 2009 to encourage strategic planning for determining suitable sites for off-shore renewable energy. Among the reforms introduced was the removal of a further requirement for off-shore renewable energy projects to obtain off-shore construction permits (Kablan and Michalak, 2012).

The integration of off-shore energy into planning processes took a further step in the UK in 2009 with the introduction of a system for preparing marine spatial plans for each UK in-shore and off-shore region under the Marine and Coastal Access Act (2009).[8] The aim of marine plans is to provide a clear and locally relevant process 'for analysing and allocating the spatial and temporal distribution of human activities in marine areas' that ensures widespread participation and input from statutory consultees,[9] experts and other stakeholders to achieve an appropriate balance between social, economic, and environmental objectives (Fletcher et al., 2013: 341). Because of the varying circumstances of different in-shore and off-shore regions, the Marine Management Organisation has divided English waters into eleven zones with the goal of producing plans for two areas every two years (Marine Management Organisation, 2012). The East in-shore and off-shore areas were the first to be selected, while planning for the South began in 2013 as part of an iterative process of 'learning-by-doing' (Carneiro, 2013).

At the time of writing, France still lacks customized legislation for marine spatial planning and, instead, utilizes other sectoral legislation and regulations to support aspects of marine planning without them yet forming into an integrated policy (SEANERGY, 2011a). Trouillet et al. (2011) describe this planning and consultation system as following a typical French planning model in which central government exerts a strong influence on local processes, using mechanisms based on those used in urban planning. It would be deceptive, however, to infer that the UK has a clearer overall plan for meeting the spatial requirements of off-shore renewables because its methodology involves developing area plans on a case-by-case basis. However, what the UK system possesses that the French system lacks is a defined *process* to determine allocations. Among the priorities for reform identified by Trouillet et al. (2011) are: promoting greater balance between decision levels to prevent excessively top-down decision-making; and better representation of stakeholders within management processes. Another emerging characteristic of the French planning system, however, is a growing influence of the regions as a result of decentralization processes initiated in the 1980s (Loughlin, 2008). However, Loughlin adds that decentralization has increased the complexity and inefficiency of French planning by creating overlapping competences between different layers of government, though the *Grenelle II* Act has attempted to make the drafting of impact assessments less bureaucratic and more participatory by allowing developers to liaise with the environmental authorities before formal authorization procedures begin and by providing for local stakeholder conferences before assessments are completed. Current planning and consenting procedures nevertheless lag behind those needed to develop the French off-shore renewables sector, leaving developers susceptible to the differing priorities of, and lack of coordination between, the various groups involved in determining planning applications (SEANERGY, 2011b).

The first more general observation to be made about the two national planning and stakeholder consultation processes concerns the absence of pre-existing legal frameworks, institutions and procedures in either country to accommodate the planning needs of off-shore renewables, and the consequent need to adapt existing frameworks and institutions until such time that bespoke institutions can be created to support and oversee the sector. Although such a situation might be expected for many climate innovations, particularly more novel ones, the process of 'learning-by-doing' and concurrent development of regulatory frameworks and technologies could be said to expose technologies to considerable risks during the early stages of development. Second, the distinctive requirements of off-shore renewables (particularly their need for physical space) have subjected off-shore renewables to a more eclectic range of interests than affect many climate innovations. This in itself may constrain the willingness of project developers and finance companies to invest in more experimental technologies. Full analysis of these constraints is beyond the scope of this chapter. However, one final feature of UK and French planning policies for off-shore renewables is the further blurring of the boundaries between CMEs and LMEs. If anything, the UK's planning systems exhibit higher levels of rules-based coordination while also seeking to embrace greater pluralism (for example, via marine spatial planning), while the French system bears stronger hallmarks of top-down, hierarchical and bureaucratic state organization. This may simply reflect the dynamic nature of the off-shore renewables sector but it also suggests an element of 'drag and lag' in the adaptation of institutional traditions to situations of rapid innovation.

Concluding discussion

The case study of off-shore renewable energy development in Europe provides a number of insights into the effects of regional and national institutions on the development and commercialization of climate innovations. The first and arguably most important requirement is to appreciate the distinctive attributes of individual types of climate innovation and their implications for the innovation processes used to promote them. Although off-shore renewables share certain features with other climate innovations (for example, lack of *a priori* market demand for their emissions reduction potential, long development lead times, and the need for government support and patient finance – see Chapter 1), the many technologies in this category also possess distinctive features that hamper generalization. Much the same can be said about the importance of understanding *institutional* complexity and variability when examining national processes of climate innovation. Typologies like VOC provide useful tools for examining characteristic features of climate innovation in LMEs and CMEs. However, the French and UK examples illustrate the need also to understand how these general factors are filtered through nationally-specific cultural, social, economic and political practices lenses to produce unique innovation processes. The further blurring of national 'traditions' as a result of EU membership and French and UK participation in collective decision-making on issues such as climate change and energy reinforces the need to understand institutional complexity and institutional diversity when examining climate innovation in any given context.

Second, the study underscores the role of governments as an important partner and source of initial momentum for many climate innovations (Etzkowitz, 2003). Although autonomous private-sector climate innovation does commonly occur in response to pre-identified market demands, deficits in specific consumer demand for emissions reduction mean that, often, government incentives are needed to spur private-sector risk-taking. In the case of off-shore renewable energy, EU intervention has ranged from general agenda setting aimed at fostering climate collectivism among its member states (for example, its involvement in international negotiations) to specific measures, such as the renewable energy directives, to establish a contractual basis for national renewable energy programmes. However, the study also found strong evidence of the influence of pre-existing political traditions influence on *how* governing authorities choose *and are able* to intervene to promote climate innovation. In particular, the study shows how the EU's emphasis on setting agendas and general goals while leaving the member states to determine how these are achieved reflects the distinctive institutional traditions and constraints created by the EU's status as a multi-level governance system and the corresponding need to develop intricate legal frameworks and institutional procedures to balance the pressures of integration with respect for the sovereignty of its member states (Jordan and Adelle, 2012).

The third major finding concerns the extent to which both the UK and France initially sought to apply standard institutional procedures to a novel socio-technological problem. In the UK, this took the form of prioritizing market enablement policies within a framework of rules governing all types of renewable energy, while France has adopted a more bureaucratic and state-centred approach whose antecedents can be traced to its nuclear energy programmes of the 1960s and 1970s. Three explanations can be proposed for this tendency: (i) *institutional path-dependency*: where governing institutions attempt to fit novel problems into existing structures and procedures on the basis of their track record of providing appropriate solutions to past problems (Jordan et al., 2003); (ii) *appearance of fit:* where existing institutional rules are presumed to be adequate if the unique requirements of climate innovations are not fully appreciated (for example, because of information or analysis deficits); and (iii) *cost-benefit evaluation:* where governments adopt a 'wait and see' approach of tweaking existing institutions and rules until more evidence based evaluations of the need for reforms to encourage individual climate innovations can be made.

However, the study also showed the continued influence of national institutional preferences on policies to support climate innovation despite defects in both the French and UK approaches to marine renewables. For example, the UK's emphasis on using markets to create *enabling* conditions over *ensuring* climate outcomes are achieved (Giddens, 2011) has theoretically promoted cost-effectiveness but has struggled to deliver patient finance (Mitchell, 2008). Conversely, French attempts to *ensure* the sector's expansion through strong state involvement in tendering, investment and spatial planning reveal a lower tolerance of uncertainty and private sector entrepreneurialism. The same factors (path dependency, appearance of fit and cost-benefit evaluations) would each appear to contribute toward explaining these continued national differences, though the evidence suggests that more static notions of path dependency (emphasizing the propensity for national institutions to follow standard operating procedures unless they are evidently dysfunctional) may play a more prominent role while uncertainty persists about the long-term viability of experimental technologies like wave and tidal power. It would be wrong, however, to stress national institutional embeddedness without also emphasizing the institutional adaptation that has taken place in France and the UK within a relatively short time period. In particular, evidence abounds of bespoke institutions and rules being created (for example, reforms in French and UK planning and the creation of the UK Marine Management Organisation) and a convergence of national approaches as a result of internal policy learning (for example, marine spatial planning in the UK) and policy transfer between jurisdictions (for example, changes in financial incentives in the UK in response to changes in Scotland and the UK's move towards feed-in tariffs).

Greater uncertainty, however, surrounds the relative capacities of CMEs and LMEs to engage in radical climate innovation. The classic hypothesis is that LMEs are more inclined to radical innovation, whereas CMEs tend to prioritize more incremental innovation processes (Akkermans et al., 2009). Certainly, the UK shows signs of frequent (and at times almost compulsive) experimentation with financial incentives, one effect of which has been to prolong uncertainty for business contemplating investment in less mature and more capital-hungry technologies (Haas et al., 2011). The UK example suggests that political imperatives emphasizing cost effectiveness and leaving markets to select technology winners have acted against market appeals for regulatory certainty. One message from the analysis, therefore, concerns the influence of *political* paradigms on climate innovation and their expression in national policies that uphold dominant paradigms rather than the imperatives of market actors, central of which is a desire for regime stability over 'optimal' policies. French renewable energy policies, meanwhile, exhibit forms of ideological 'lock in' based around state investment and oversight, and a more cautious approach to market instruments that appears less innovative but which may prove more effective in encouraging corporations to innovate. As such, LME and CME categories may hold less explanatory power than broader notions of political ideology when seeking to understand national variations in climate innovation (Taylor, 2004).

Finally, the case of off-shore renewable energy underlines that certain types of technological innovation require not just political and market acceptance, but also the building of societal acceptance through increased engagement with stakeholder groups affected by new technologies and practices (Wüstenhagen et al., 2007). Put another way, new innovation 'rules' (such as marine spatial planning) are needed to accommodate the distinctive features and geographies of off-shore renewable energy. However, the French and UK cases show that such frameworks are still deeply informed by states' underlying *democratic* traditions that again extend beyond their particular brands of market capitalism.

In some respects, it is unsurprising that VOC thinking provides only partial explanations for why countries adopt different approaches to climate innovation. Almost by definition, new problems like climate change require experimentation and going beyond existing 'comfort zones'. Equally, it is unrealistic to expect national traditions to be jettisoned when they have repeatedly been judged to be proficient in the contexts in which they operate. The case of off-shore renewable energy underlines the need to look at factors that extend beyond VOC in order to understand more fully the institutional complexities shaping the development and uptake of climate innovations.

Notes

1 I acknowledge financial assistance provided by Intelligent Energy Europe (Streamlining of Ocean Wave Farm Impacts Assessment, SOWFIA: IEE/09/809/SI2.558291) and INTERREG IV A (Marine Energy in Far Peripheral and Island Communities, MERiFIC: 3726/4122) for this research. All views expressed are the author's own and do not represent the opinions of either organization. I also thank Peter Connor, Angus Vantoch-Wood, Jiska de Groot and Yannis Kablan for their contributions to the research underpinning this chapter.
2 The European Commission has the exclusive right to propose legislation and the responsibility to ensure the member states meet their treaty obligations, while the Council and Parliament have co-equal legislative authority to accept or reject Commission proposals for legislation. The European Court of Justice is the authoritative arbiter on the interpretation of the EU treaties and specific pieces of EU legislation (Jordan and Adelle, 2012).
3 The directive states that states should achieve 20 percent of their 2020 target by 2011–2012; 30 percent by 2013–2014; 45 percent by 2015–2016; and 65 percent by 2017–2018 (Official Journal of the European Union, 2009).
4 Stephanie Merry, Head of Marine at the UK's Renewable Energy Association, at the opening plenary of the 2012 SeaTech Week in Brest, France, October 2012.
5 It is important to bear in mind that such characterizations can be problematic where mismatches occur between stereotypical images and realities (Szarka, 2008).
6 Scotland operates a parallel scheme, the Renewables Obligation Scotland.
7 The marine public domain is the property of the state and comprises all shores, seabed and sub-soils up to twelve nautical miles from the mean low-water mark. This does not cover the French Exclusive Economic Zone, which covers up to 200 nautical miles from the coast where a large portion of wind and wave energy devices might be located (SEANERGY, 2011a).
8 Scottish seas are covered by the Marine (Scotland) Act 2010.
9 For example, the Environment Agency, English Heritage, Natural England, Health and Safety Executive, and the Centre for Environment, Fisheries and Aquaculture Science.

11
Conclusion: A Way Forward

Neil E. Harrison and John Mikler

This book is founded on three premises. First, we accept the scientific consensus that the Earth's climate is changing due to global warming caused by human activities that have resulted in too many greenhouse gas (GHG) emissions from the combustion of fossil fuels. Climate change should, if possible, be mitigated. Secondly, we assume that the countries that are the major emitters of GHGs will continue to organize their political economies as some distinct variety of capitalism, and that liberal capitalism is one of these. All states, including liberal capitalist states, cannot easily choose to alter the institutions that underpin their variety of capitalism. Thirdly, we accept that the governments of these states would more readily mitigate climate change if they could avoid regulating social activity or harming economic growth. In short, the puzzle this book has addressed is how to mitigate dangerous climate change within a capitalist economy without significantly changing the socio-economic system or retarding social welfare.[1]

In pursuit of this puzzle, we have primarily focused on the United States (US) as the archetypical liberal market economy, and the world's second largest contributor to GHG emissions after China. Because of its size and importance, the US is the potential leader for a negotiated international response to the climate change challenge and for solutions to mitigate climate change in other capitalist states. As the US goes, so often has the world, and in this case particularly those states whose variety of capitalism most resembles its own. Despite Congressional opposition, President Obama has taken some small steps toward this leadership role for the US. Under the 2009 Copenhagen Accord, the US agreed to reduce its GHG emissions by 17 percent from 2005 levels by 2020. In 2009, the Obama

Administration began to fund nascent radical energy-saving technologies through the Advanced Research Project Agency-Energy (ARPA-E) program managed by the Department of Energy, and in 2011, the US helped to launch negotiations for a replacement of the Kyoto Protocol that would include leading developing nations. However, the evidence to date, as summarized in Chapter 6, does not support the contention of some observers that the US will reach its target (see also Burtraw and Woerman, 2012).[2]

The US would be most likely able to lead international negotiations as President Obama wishes, if it uses technological innovation to reduce its GHG emissions and avoid intrusive regulations that would result in wrenching social or economic change. This is also the case for other liberal capitalist states where a preference for less, or more accurately more 'arm's length', government intervention in markets and society prevails. Full uptake of available technologies that mitigate emissions is unlikely in the foreseeable future, not least because of the sunk costs in the fossil energy infrastructure. However, even if that were possible it would not prevent dangerous climate change. That requires new technologies, probably radically designed for the purpose of reducing GHG emissions, to be developed and diffused through national economies, and throughout the global economy. In addition, the location and design of current technologies may mean that developed countries are using high cost solutions that are only marginally effective because they are not designed specifically to reduce GHG emissions (Helm, 2012). Emissions reductions caused by products and production processes from market innovation are peripheral to their designed purpose and, based on the evidence to date, inadequate. Thus, in Chapter 1 we argued that the process of 'climate innovation' – purposeful technological innovation to mitigate climate change – is markedly different from the market innovation process that has been more widely studied and is, therefore, better understood. Climate innovation cannot rely on market forces alone because mitigation of climate change is a non-market objective. Market innovation responds to market demands because it is driven by capitalist accumulation, but climate innovation must be driven by the environmental objective of avoiding dangerous climate change.

In this chapter we first summarize how the results of the research in this book support and elaborate the theoretical framework we developed in Chapter 2. This revolved around four dimensions needed to support climate innovation: patient finance; management autonomy; acceptance of uncertainty; and 'climate collectivism'. Three of the four dimensions are visible within the US national innovation system, and by implication the innovation systems of other liberal capitalist states. The fourth is conspicuous by its absence. We then show why the most liberal capitalist states will continue to fail to generate sufficient technological innovation. Without substantial and *unanticipated* changes, US governments will be unable to rectify the innate deficiencies in private sector research, development, and dissemination of new technologies designed to mitigate GHG emissions. We also expect most other liberal capitalist states' national innovation systems to fail to generate sufficient climate innovation and doubt that coordinated market economies will be able to fill that gap. We therefore conclude that it is highly unlikely that sufficient technological innovation will prevent dangerous climate change without substantially disrupting the economy or society in liberal capitalist states.

Testing the four institutional dimensions

We argued in Chapter 2 that some measure of all four dimensions would need to be evident for a nation's institutional structure to generate climate innovation. The research presented in this book primarily speaks to three of the institutional dimensions developed in Chapter 2. It shows that the innovation system of the US displays aspects of patient finance, uncertainty acceptance, and management autonomy though there are qualitative deficiencies in each. The fourth, climate collectivism, is essentially absent. Inclusion of this dimension in our theoretical framework is supported by research and was intended to capture a potential cause of the rapid technological progress of the emerging East Asian nations such as China, South Korea and Singapore. This, we believed, would make the theoretical framework applicable to innovation systems across the whole range of national varieties of capitalism and local histories. But in Table 2.3 we suggested that climate collectivism would be 'low' for the US, and we are not surprised that none of the chapters in this book found any evidence of this institutional dimension. Drawing on the authors' analysis and findings we discuss below why this is the case, as well as why institutional change in respect of the other three is required to support the climate innovation necessary.

Patient finance

The radical climate innovations necessary to mitigate climate change demand financial support that is patient enough to support the long-term research and development (R&D), especially the science, demonstration, and pre-commercial (or 'scale-up') stages of the innovation process. In Chapter 2, we argued that capital markets are often shortsighted and that investors are increasingly unwilling to play a 'long game', resulting in under-investment in paradigm-changing innovations. Some investment is more patient. For example, once market viability can be demonstrated, venture capital is able to justify investment in the roll-out and marketing of a finished (or nearly complete) product.[3] Yet, even angel investors – early stage high net worth investors – only hold their investments for an average of 3.67 years (DeGennaro et al., 2010). In addition, markets commonly undervalue companies that use R&D as a critical component of their strategy, are more R&D intensive than the average in their industry, and invest in basic science research, all of which are components of climate innovation (Ciftci et al., 2011). Finally, equity capital is supposed to provide a stable financial foundation for the long-term evolution of a business but it is increasingly a costly alternative to debt capital (Carter, 2013). As the benefits of long-term strategies and innovation investments are undervalued and the dominance of high-frequency trading (driven by technological advances) increases, capital markets become increasingly short term and unsupportive of the radical innovations that can generate rapid mitigation of climate change together with economic growth. In fact, the average holding period for stocks and shares in the US fell from four years in 1945, to eight months in 2000 and two months in 2008 (Patterson, 2012). By some estimates the average holding period is now just *twenty-two seconds* (Turbeville, 2013).

The short-termism promoted by capital markets is discernible for the large US public companies reporting to the Carbon Disclosure Project (CDP), and in the interviews conducted with Australian corporate representatives interviewed from companies listed on the Dow Jones Sustainability Index (see Chapter 6). Although these companies have substantial internal financial resources and ready access to external capital, few admitted to investing in long-term R&D projects with uncertain pay-offs, or had fully integrated climate change into their long-term strategy. The majority of the companies reporting to CDP were investing in short-term projects to reduce their energy intensity that generate rapid returns to investment in the form of reduced costs and higher profits. Only a third of the energy efficiency or product development projects reported by forty-two companies had a pay-off beyond three years, and very few companies had explicitly incorporated climate change into their long-term strategic plans or foresaw opportunities in developing products or services specifically to mitigate

GHG emissions. Both the CDP reports and the Australian interviewees emphasized that until there is a quite explicit and imminent 'business case' for climate mitigating products or services, companies *cannot* invest in them. In other words, short-term financial considerations trump longer-term climate innovation.

Any concern companies have about market uncertainties are only exacerbated by uncertainties about the future institutional environment, and in this respect regulatory uncertainty also reduces opportunities for external financing of climate innovation. Long-term financing is available to credit worthy organizations – the US Treasury has returned to issuing thirty-year bonds and many US municipalities are able to issue ten-year bonds at reasonable rates – as long as interest payments are substantially assured. However, uncertainty about how future rules may frame the market adds non-market risks to the competitive risks in the market, making financial return even more unpredictable and less attractive to investors. Thus, regulatory uncertainty not only exacerbates the natural short-termism of capital markets, but also stalls long-term, and probably even medium-term, corporate strategies.

The current regulatory uncertainty is especially problematic for industries that naturally have long-term investment horizons. For example, in their CDP reports US utility companies expressed great concern. Within the relatively more stable regulatory framework of recent decades, regulated utility companies have been able to raise the funds necessary for power plant, pipeline, or grid construction by offering investors a high yield that is essentially guaranteed by the regulatory authorities that set local electricity prices. Unable to predict electricity production constraints or demand, they have resorted to smaller projects aimed at conserving energy and reducing demand, obviating the need to choose among power plant technologies. Offshore wind energy faces a similar challenge – a large capital investment with a long-term payoff. In Chapter 10, Ian Bailey shows that the UK, the European Union's (EU) most liberal capitalist economy, has changed its regulations and financing terms more frequently than France, and that this has tended to deter private investment in offshore renewable energy.

It is therefore not surprising that high-risk investments in R&D have had to be increasingly supported by government programs in recent years. As detailed in Chapter 3, the range of programs in the US now amount to a 'powerful and proactive' developmental network state (DNS) that calls into question the ideological insistence that US innovation relies, and should rely, exclusively on market forces. Yet, as Robert MacNeil demonstrates, the generic research support mechanisms that have driven innovation in industry sectors such as information technology are not as applicable to climate innovation because they were not designed to create markets based on non-market prerogatives, and they do not support the kind of long-term science-based research that climate innovation requires. The DNS has been effective at driving market innovation because it funds the early stages of products and services intended to satisfy nascent market demand. The research support, technology transfer, and networking it provides fill gaps in the normal market innovation process. That process, as we have argued, is designed to sell goods and services to deliver profits. But climate innovation needs patient financing more targeted to the non-market goal of mitigating climate change as rapidly as possible with the minimum of social or economic dislocation. To be wholly effective climate innovation should be supported across much of the economy and especially in industrial sectors that consume energy such as agriculture, airlines, residential housing, commercial construction, manufacturing, and chemicals.[4] However, MacNeil shows that while the US federal government's elaborate DNS has been effective in stimulating many technologies, it has failed to sufficiently advance clean energy technology. Despite decades of effort its only significant success – that of 'fracking' – has merely served to further entrench fossil fuel energy that already enjoys price, infrastructure, and customer familiarity advantages, while not addressing the relative lack of effective demand for *clean* energy.

The obvious solution of creating such demand through regulations to increase the relative cost of fossil energy has been stymied by a powerful fossil fuel energy lobby and the continuing Congressional opposition to energy regulation motivated by a neoliberal ideology. MacNeil concludes that the failure of the DNS to support clean energy technologies to commercialization is a consequence of its purpose. It was established to help companies innovate for the market and to increase economic growth. It was not designed for social or political goals like avoidance of dangerous climate change. MacNeil therefore recommends a progressive move away from a neoliberal ideology in Congress or appointment of 'a powerful institution within the federal government that is largely immune from the political impediments' raised by neoliberalism as the focal point for climate innovation.

Sufficient political movement away from liberal capitalism is not likely in the foreseeable future though. The coal and fossil fuel energy states that are generally rural and predominantly Republican have sufficient power to block any Senate action to mitigate climate change. Robert MacNeil suggests that the Department of Defense's (DOD) Defense Advanced Research Projects Agency (DARPA) may be an appropriately non-political institution from which to direct economy-wide climate innovation. However, despite the DOD's determination that 'climate change and energy are two key issues that will play a significant role in shaping the future security environment' (US Department of Defense, 2010: 84), DARPA is neither sufficiently market-oriented nor is it intended to be climate-oriented. In addition, a Congress focused on budget-cutting has recently taken its knife to DOD funding suggesting that defense is no longer relatively immune to politics.[5] While a non-ideological institution is needed to lead and coordinate climate innovation, the DOD seems an unlikely candidate.

Modeled on DARPA, ARPA-E attempts to fill the gap between basic research that is already well funded and late stage product development that venture capital is well able to support (LaMonica, 2012). However, ARPA-E is underfunded and limited to applied research in search of technologies to reduce energy imports or emissions from energy use. If political ideology would allow, a much more substantially funded 'ARPA-CC' (Advanced Research Project Agency-Climate Change) could apply the same strict scientific and financial review of applied science projects that ARPA-E has developed across much of the economy. ARPA-CC would select projects based on their projected reduction in GHG emissions and would be agnostic about the industrial sector that the proposed innovation would affect. More generally, we suggest that a large source of funding to provide patient capital in a liberal capitalist economic setting where this is increasingly lacking would be required. And it must be to fund long-term climate innovation specifically, not market-led innovation. Needless to say, this is a daunting prospect given the ideological and institutional setting in which this must occur.

Management autonomy

Managers, Boards and many shareholders jealously guard management autonomy, though for different reasons. Managers demand the freedom to choose and implement their plans as they see fit and Boards of Directors protect the autonomy of their managers from external influence to increase their own power and influence over them. Shareholders are primarily concerned with the ability of the managers they have 'hired' to generate a healthy return on their investment.

They desire managers who are free to act on this imperative, though some would like these returns delivered with socially and environmentally sound business strategies. Management autonomy is less of a concern for institutional shareholders, who hold their investments longer and take a more active interest in the operations of the company beyond its ability to deliver short-term returns on their investment. However, as noted in the previous section, such longer-term shareholdings have become less and less a feature of liberal capitalism as practiced in the US.

In Chapter 2, we argued that US corporate governance rules give managers more autonomy than managers in other countries, although this ideal is shared with other liberal market economies at least in spirit if not necessarily to the same degree in practice. It is evident from the research in the chapters that this is certainly true in the context of climate change. David Levy and Sandra Rothenberg explain in Chapter 5 how companies 'learn' about environmental issues. Theirs is a fascinating study of how two companies in the same industry perceived climate change quite differently before it became broadly accepted as settled science in the business community. But it also shows that the corporate response to this important environmental issue was wholly determined within the existing managerial structure. There is no evidence that other stakeholders were involved in their strategic decisions on how to comprehend and respond to scientific uncertainty. They relied on 'their' corporate scientists to interpret and translate the climate change scientific research within their corporate institutional settings.

Our review of CDP reports in Chapter 6 shows that companies determine their climate change strategies and investments wholly within management; that managers are offered incentives – usually financial – to reduce emissions; and that companies with divisional or business unit structures often delegate decisions on climate projects to divisional committees. Our Australian corporate interviews show that investments in climate mitigation are held to the same demands for financial returns as any other investment. This means that decisions are internal, wholly within management, and geared to financial metrics like revenue and profit streams, about which the investor community is most concerned. The CDP reports and interviewees implicitly, and often explicitly, state that climate projects are selected by management for their contribution to a business case for short- to medium-term profits.

In Chapter 7, Dimitris Stevis argues that the US political economy has become more liberal in recent decades and the balance of power between capital and labor has shifted. In many sectors of the economy managers are almost completely free of the 'burden' of negotiating with unions over compensation and benefits and, unlike in some European countries, are able to wholly ignore the needs of labor in making investment decisions. Unions have responded in a variety of ways from organizing in previously overlooked sectors to merging across sectors, breaking the mould of industry or sector unions common in the US. A few unions have co-opted environmental groups in an effort to tilt the scales back toward labor by drawing the state into labor relations through a program of environmental conservation that includes climate innovation. Their hope is that the range of managers' choices will be limited by targeted regulation in selected industries that would aid environmental conservation while creating many good quality 'green' jobs. While this effort has yet to produce significant results, even with a Democratic occupant of the White House, it does suggest that social and political action could leverage climate innovation to change managers' calculations and include non-financial considerations in their decision-making.

What holds at the national level appears to be projected internationally. In Chapter 8, Jeffrey McGee essentially agrees with Stevis that in recent decades a strong streak of neoliberalism has appeared in the US political economy that has been most clearly revealed in its policy toward international climate change negotiations. US support for market-based mechanisms reflects this ideological, and institutional, bias. While he does not make the argument explicitly, the implication is that, as Friedman (1970) asserted, in the US the business of business is to make profits, and that managers' responsibilities are primarily to their shareholders. This in turn implies that managers should be autonomous except for the financial demands of shareholders. By comparison, China lies at the other end of the spectrum that McGee draws in Figure 8.1. As John Mikler and Hinrich Voss show in Chapter 9, while US corporations remain focused on the bottom line and market imperatives, managers in Chinese MNCs remain responsible, at least in their overtly stated motivations in company reports, to the Chinese state even when operating abroad. This is an unusual state of affairs for purportedly capitalist companies, but it demonstrates our point that actors such as these MNCs are subject to the rules established by the institutional frameworks in which they are embedded.

US corporate governance rules allow managers to act autonomously with little concern for external influences. Because most shareholders act less as owners than speculators – with their increasingly short average holding period – managers can largely ignore the needs of their purported employers. Managers' expressed pursuit of revenues, profits, and share price appreciation may be as much the result of their share-based compensation as their concern for the interests of shareholders. Yet, there is little evidence from the research reported in this book that more than a minority of managers (or the Boards they serve) want to use that autonomy to invest in climate innovation as we have defined it. In Chapter 2, we argued that the autonomy of managers should permit them to take the long view and invest in research and development for radical innovations. That this is not occurring is less caused by autonomy than by the mental models of managers. Currently, they are trained in business schools to pursue financial metrics, and are compensated by share price-related bonuses and fêted in the financial press for meeting them. This is a consequence of a liberal political economy. There are two principal ways to change their minds: regulation or persuasion.

Realistically, regulation does not change mental models but only the behaviors that emanate from them. As the rules change, managers would pursue the same goals by different means. Social pressure, however, may change managers' minds and goals and, thus, their behavior. The Corporate Social Responsibility (CSR) movement is a salient example, and the CDP process is a similar effort to insert climate change into corporate governance by making managers cognizant of their effects on the climate and their options for mitigating their company's emissions. Indeed, several ideas fit together to pressure changes in managerial behavior: the urgency of climate change that they intellectually recognize sets the goal of mitigation; CSR instills the ethical concepts in their mental models to make them aware that they owe a duty to the society beyond their pocket book, the business, and shareholders; stakeholder theory offers a business argument for a broader duty of care; and corporate accountability theory argues for a reckoning of the actions for which they are held responsible (Wilson, 2003).

But is this enough? If managers continue to use their autonomy largely to pursue narrow financial goals, external limits may need to be placed on their choices. Government could mandate changes in corporate governance rules to reduce their autonomy and make managers more responsive to a broader array of stakeholders. While no action of this kind is in the offing, even the threat may open up manager's perception of their duties. Governments might also require companies to explicitly include climate change mitigation in their corporate purpose and processes. A carbon price on GHG emissions or a cap-and-trade system may also be part of the solution, although Australia's carbon tax looked set to be removed as this book went into print, and the European experience with a cap-and-trade system has been less than a roaring success as the price of GHG emissions has plummeted since it was introduced. But more than this, both encourage a *market* view of the problem of climate change rather than an environmental one, and as such they underpin, rather than reform, the financial return mentality and motivations of managers, boards and their shareholders. Even if such market measures encourage firms to bear a cost, this does not mean that they will endogenize a climate change mitigation mentality.

Uncertainty acceptance

As discussed in Chapter 2, US success in innovation is sometimes ascribed to its cultural acceptance of uncertainty. A central strength of markets is that they encourage a range of adaptive innovations, or *mutations*, in response to changes in the external environment (Nelson and Winter, 1982). The market 'selects' among these mutations. This selection is exogenous to any individual company but is endogenous to the economic system, and the uncertainty that the selection process produces is accepted as a common risk of doing business. Changes in the political environment, however, are exogenous to the economic system and are challenges of a different order as they produce risks from non-market events that are not predictable in the ordinary course of business.

The DNS described in detail by Robert MacNeil in Chapter 3 is designed to temper market risk, that is, the risk that firms' investment choices will not be selected by the market. It is a sophisticated adjunct to the substantial investment that the US government makes in basic science, but it is not designed to meet GHG emission reduction goals (see also Mikler and Harrison, 2012). It relies on the logic of serendipity that liberal market economies are heir to that, if enough technological innovations are generated, eventually they will add up to an effective climate mitigation strategy. But it does nothing to assuage regulatory uncertainty that, as we noted in Chapter 6, appears to be constraining investment in climate innovation. The companies reporting to the CDP are leaders in climate change recognition, emissions measurement, and response strategies. Yet, most of these companies are investing only in short-term climate mitigation projects primarily designed to reduce energy intensity, reduce costs and increase productivity – a consequence of a liberal capitalist mindset and the uncertainty of the regulatory future.

Chapter 5, David Levy and Sandra Rothenberg show how companies handle non-market uncertainties. Even with the best research, market intelligence, and focus groups companies can always produce a Ford Edsel (Bonsall, 2002). But companies are familiar with market uncertainties and the successful ones have developed the organization and processes to manage them. Climate change presents them with uncertainty that is much less amenable to quantitative research and analysis. In their choices on how to respond to the claims of incomplete climate science, Ford and General Motors (GM) interpreted this non-market uncertainty through the frame of their organizational structure and product range. Ford primarily saw risks and joined a campaign to combat the message of the climate scientists. GM, however, saw opportunity in the regulations and changes in consumer behavior that climate change might engender and was more readily persuaded by the science. Levy and Rothenberg focus on the role of a senior GM scientist who became part of the epistemic community of scientists pushing for political action (Haas, 1992). She was able to span the boundary of the organization and bring an understanding and acceptance of the science to senior management in large part because GM was working on electric cars and hybrid drives. While her role was important, the scientific and policy uncertainty that climate change produced was interpreted by GM management as an opportunity, but by Ford as a risk. We saw much the same pattern in the CDP reports in Chapter 6. The companies with a diversified product base or geographically diversified operations were more sanguine about regulatory uncertainty, apparently believing that they will be able to find market demands to satisfy whatever form of climate regulation is eventually selected.

As we concluded in Chapter 6 business needs a stable framework within which to consider the business case for climate innovation. Market uncertainty it can manage; uncertainty about the rules and regulations surrounding the market not only are less familiar and manageable but also exacerbates the natural tendency of liberal market economies to gravitate to short-term returns on investment even while longer-term investments may be more rewarding. Nature abhors a vacuum. Business detests regulatory uncertainty. Thus, it is important both for economic growth and climate change mitigation that a stable institutional framework be constructed to move market activity towards emissions reductions and entice more investment in climate innovation as soon as possible. Indeed, it may be more important to reduce regulatory uncertainty than to perfect the mechanism by which emissions are reduced.

Climate collectivism

As we commented above, none of the research in this book shows any evidence in the US of climate collectivism as described in Chapter 2. Because of the special needs of climate innovation, some measure of collectivism probably is necessary, albeit not necessarily sufficient. We do not mean that only more coordinated forms of capitalism, such as exemplified by the East Asian developmental states, or Chinese state-led capitalism, or German corporatism can generate climate innovation, but it is certainly the case that a high degree of national cohesion is required to address such a 'diabolical' collective action problem (Garnaut, 2008). After all, as the US space program demonstrated, non-economic goals in particular require institutional structures that draw together the resources and skills of profit-seeking companies but organize them for the non-economic national purpose. Today, private companies like Space-X are now active participants in the US space program. Similarly, generating sufficient advanced technology to mitigate sufficiently GHG emissions and climate change requires more unity of purpose than the mass of profit-seeking companies organized simply through the market can achieve.

Depending on one's political perspective, markets either are able to generate such a large number of innovations that any social or environmental problem is soluble, or markets fail at this task because they are infested with insoluble collective action problems (Hayek, 1994[1944]; Olson, 1965). The DNS hidden within the US political economy is an example of the former kind of thinking. It is a scatter-gun approach to technological innovation that has been effective because it is market-oriented. It has no social purpose beyond abetting capital accumulation as broadly as possible, which, of course, is a consequence of the liberal capitalism that underpins the US political economy. Relying on the DNS to drive climate innovation is a salient example of the US's inability to organize climate collectivism. Clearly, we adhere more to the latter perspective, as even companies that appear to agree cannot act together. The CDP reporting companies we analyzed in Chapter 6 represent nearly two trillion dollars of revenue between them. They almost unanimously recognized the reality of climate change, expressed concern at the risks of regulatory uncertainty, and therefore reported that they were lobbying the US Congress to reduce these risks. Yet, Congress has not acted and business has contributed much less to mitigation targets than needed to meet the US's 2020 target. Acting individually, most CDP reporting companies set inadequate targets and, despite these apparently common interests, they have been unable to act collectively to eliminate the regulatory uncertainty that they fear.[6] In a similar vein, as discussed by Dimitris Stevis in Chapter 7, even with the support of unions through the BlueGreen Alliance, business has been unable to achieve its desires.

Within the context of national liberal capitalism, in Chapter 4 Stratis Giannakouros and Dimitris Stevis show that the State of Colorado under Governor Bill Ritter sought to create some local collectivism with an 'ecosystem' of new institutions and regulations. Founded through a constitutional amendment, the ecosystem was a 'middle way' around which most Coloradans of various political stripes could rally. It increased regulation on the fossil energy industry and required the state's regulated utility to reduce environmental damage from oil and gas drilling while creating demand for renewable energy manufacturing which then moved into the state. Yet, before the ecosystem could take root a new governor (an alumnus of the oil and gas industry) began to dismantle it and eviscerate its nascently forming institutions. This was relatively easy to achieve, because the national liberal institutional context did not support such a sub-national initiative. There are two key lessons in this tale. First, that effective climate mitigation can be coupled with economic development only through a comprehensive array of policies based on a foundation of institutions designed for the purpose. Second, without a *national*, collective basis for action laudable sub-national initiatives, that many expect ultimately to produce a national response, may fail to deliver what is promised. The American Recovery and Reinvestment Act of 2009 conspicuously failed to support Colorado's nascent institutions and routed most climate investment directly to Washington's preferred targets within the state. While sub-national approaches in the US federal system may lead to or circumvent a national response to climate change, as they did for ozone depletion, they are fragile without national support underpinned by some national measure of climate collectivism (Byrne et al., 2007).

It would appear that without an existential crisis the US, and liberal capitalist states like it, will continue to eschew collective action to mitigate climate change in favor of the market freedoms that cannot solve the problem and which, we would argue, have helped to create it.

A way forward

Giannakouros and Stevis's aforementioned analysis in Chapter 4 does, however, demonstrate the importance of constructing an institutional framework within which to generate climate and other clean technologies. This is why Governor Ritter described his New Energy Economy as an 'ecosystem' designed to protect the environment while promoting economic development. Ultimately, governance is always about constructing ecosystems within which individuals and organization pursue their lives and interests. Survival within an ecosystem is determined by the ability of an organism to adapt to the rules of the system which determine which behaviors are successful (Kauffman, 1995; Levin, 1999).

Social systems operate similarly except that governments attempt to make complex social ecosystems teleological, to meet specific social goals (Nelson and Winter, 1982; Sprout and Sprout, 1971). They use policies to 'fiddle' with the parameters of the system to achieve specific local outcomes, a small reduction in poverty or an increase in wind energy production (Meadows, 2008). Yet, to really change a system requires much more than fiddling as the planet burns. It requires a change in the system's goals or the paradigm that drives it. The paradigm of natural ecosystems is survival within their environment; the paradigm of the US political economy is neoliberalism, as Jeffrey McGee argued in Chapter 8, or liberal capitalism more generally as we have named it.[7] Legislation from Congress, regulations issued by the Executive branch, and the structure and form of market activity are a consequence of the institutional framework that has evolved over time driven by a liberal capitalist system paradigm.

Changing the system paradigm to take account of environmental concerns such as climate change is very difficult. Mikler (2009, 2010, 2011) has previously argued that it is unlikely, and that essentially the more things change the more they stay the same. Or to put it more technically, the path dependence produced by the historical embedding of institutions means that they tend to endure. Institutional *bricolage* is more likely than paradigm shifts. Harrison (2014) is thankfully more optimistic that while difficult it is not impossible to change the system paradigm over time. Even so, we both believe that the system paradigm of liberal capitalism will not change within the foreseeable future. For example, despite the efforts of the Blue-Green Alliance of unions, environmentalists and some companies investigated by Dimitris Stevis in Chapter 7, we doubt whether this creative political amalgam of labor and environment with support from some business interests will be able to draw in the federal state, or sufficient sub-federal states, to change the institutional context of labor relations as the BlueGreen Alliance desires. We agree with the Alliance that changing institutions is preferable to experimenting with policies; because institutions are difficult to change, firms' regulatory uncertainty is reduced. Yet, the Alliance has won no battles in what must be a long war.

Therefore, the US will continue as the exemplar of liberal capitalism, relying on only three of the four dimensions of climate innovation, and dependent for its climate innovation on the few more enlightened managements who can look beyond short-term financial gain and avoid a simplistic response to the incentives embedded in its institutions. The evidence from the research presented in this book shows that it is weak on patient finance and privileges short-term investment returns; that corporate managers have sufficient autonomy to take risks on long-term innovations but usually reject them in favor of short-term returns; and that the ability of US firms to accept market uncertainties does not translate into a willingness or competence to embrace regulatory uncertainty. Even supposing that the US could correct its deficiencies in these three institutional dimensions – such as by instituting an ARPA-CC; broadening the range of stakeholders managers choose to serve; and framing a stable institutional framework to support climate mitigation – the world cannot look to the US to lead in climate change mitigation. Without climate innovation the US will be unable to reduce its GHG emissions fast enough to be accepted in the global leadership role that President Obama has proclaimed as his goal.

Where will that leadership in climate innovation and negotiations come from? In Chapter 10, Ian Bailey describes in detail the substantial differences in policies for encouraging offshore renewable energy in the UK and France. This is a fascinating story on several levels. For example, as the most liberal capitalist member state of the EU, the UK engaged local interests with a cohesive strategy but changed the subsidies, which are needed to make projects viable, several times. Meanwhile France hewing to its more centralized *dirigiste* system of economic management counted local concerns for little but prepared a more stable national regime of regulations and financial support. The French approach seems to have been more effective. The attempts of these neighboring nations to fit these new technologies into their unique and long-evolved institutional structures only serves to highlight the difference in the designs for offshore renewable energy and is a case study of an essential weakness in the EU project. As Bailey explains, it is an 'intertwined and evolving mélange of institutional contexts that reflects the EU's distinctive political arrangements' in which 'member states retain near-exclusive control of how agreed objectives are achieved'.

Perhaps surprisingly, this 'multi-level governance' process allowed for 'mutual leadership reinforcement' and an ideational competition among the states, the European Commission, and the European Parliament that has been central to the EU's international leadership in international negotiations (Schreurs and Tiberghien, 2007). For the decade after the US Senate rejected the Kyoto Protocol, the EU effectively led the UNFCCC negotiations. Despite large differences in national targets and abilities and the need to reach consensus on an EU target, at Kyoto the then fifteen members of the EU collectively agreed to reduce GHG emissions in the 2008–2012 period by 8 percent below the 1990 level. A burden-sharing arrangement based on the relative wealth of each member at the time of the agreement allowed national targets to range from an 18 percent cut in Luxembourg and an 11 percent cut in Germany to a 27 percent increase for Portugal.[8] The aggressive GHG emission reduction targets that the EU has consistently adopted show a degree of climate collectivism showing through its regionally complex decision processes buttressed by public support (Schreurs and Tiberghien, 2007).

For the next decade, EU leadership of the negotiations was built on offering many new ideas, increasing its own emissions reduction target, and constantly opposing the US and its Umbrella Group allies (Australia and Japan), especially during the US's period of neoliberal foreign policy that Jeffrey McGee describes in Chapter 8. The EU has implemented the world's first cross-national Emissions Trading System that covers some 45 percent of EU emissions; in March 2007 the EU unilaterally committed to reduce its emissions by 20 percent by 2020 from 2005 levels; and by 2011 the fifteen members of the EU who had ratified the Kyoto Protocol had reduced their emissions by 15 percent, nearly double their initial target. In Mikler and Harrison (2012), we concluded that cooperative market economies, of which many larger EU nations are examples and which authors such as Drezner (2007) argue the EU itself resembles, have a climate innovation advantage. This rapid reduction in emissions during a period of economic growth may be evidence of this even if the advantage was driven by heavy subsidies for renewable energy systems.

Yet, the EU has lost its leadership position. By 2010 the EU appeared on track to its 2020 target of 12.7 percent renewables but a detailed analysis of states' progress concluded that 'many Member States will need further measures to ensure the achievement of their targets' (European Commission, 2013). As Schreurs and Tiberghien (2007) anticipated, enlargement (to today's twenty-eight members) has increased the diversity of economic strengths and environmental capacities among the members. In addition, the windfall one-off GHG emissions reductions from the unification of Germany and the UK's switch from coal to gas are not likely to recur. The economic woes in the Euro area and in the UK also have stifled EU ambitions to set more extensive emissions reductions. Finally, a changing dynamic in the international negotiations was already evident at Copenhagen in 2009. The EU-US competition has been replaced by a developed-developing world divide (Bodansky, 2010).

This appropriately portends a shift in leadership of the UNFCCC negotiations. President George W. Bush argued that his refusal for the US to actively participate in the negotiations was driven by China's opposition to targets and timetables. China and the rest of the developing world used to see GHG emissions targets as a brake on their development. But by 2010 China and India had adopted carbon intensity targets. China pledged to reduce its emissions per unit of GDP by 40–45 percent by 2020 from 2005 levels (Bodansky, 2010), and China's breakneck growth has caused many environmental problems and, pressed by social discontent, it is responding with massive programs to reduce air pollution in many northern China cities (Wee and Popeski, 2013).

Driven by such needs China *could* become a leader in climate innovation. It would appear to score well on all four dimensions of climate innovation. The Chinese Communist Party could provide patient finance for long-term investment in radical technologies to mitigate climate change and develop a stable institutional framework to mitigate private sector risks in R&D investment. As John Mikler and Hinrich Voss show in Chapter 9, Chinese capitalist companies are closely attuned to the needs and demands of their home state. Therefore, we can expect that formally private companies and investors will quickly and appropriately respond if there were a state pronouncement that China intends to take a lead in developing the technologies that the world is going to need to mitigate dangerous climate change. Coupled with rising concern for the environment among educated Chinese, this means that climate collectivism, we would predict, is largely assured should this occur. Because climate innovation requires basic scientific research and applied science in a wide range of domains from materials to energy, a Chinese national goal of climate mitigation through technology may also be an appropriate way to drive forward research, higher education, and the economy.

Now, this remains speculation based on our four dimensional theoretical framework of climate innovation. Even so, we and the contributors to this volume have demonstrated that liberal capitalism falls short of mitigating climate change through climate innovation. On the basis of Ian Bailey's analysis it appears that the EU will fare little better. John Mikler and Hinrich Voss's analysis suggests that China might be able to produce the climate innovations that the developed world will need (and which it could buy) to mitigate their emissions and protect their populations. Yet, one thing is clear, more research is needed particularly in the EU and China to unravel in more detail the potential for climate innovation specifically in non-liberal states. Otherwise, if no nation steps up to press forward with climate innovation the world may be left no option but to 'live with' and adapt to whatever climatic changes result from a human failure to restrain ourselves and prevent dangerous climate change.

Some scholars and governments are already anticipating this failure and researching how we may adapt to climate change (Adger et al., 2010; Moser and Boykoff, 2013; National Research Council, 2010; Pelling, 2011; Schoon, 2012). Certainly, when a market for climate adaptation emerges liberal capitalism will come into its own, but we still hold out hope that governments will see reason and adjust their institutions to rapidly expand private investment in climate innovation in advance of the predicted catastrophes of dangerous climate change. As Winston Churchill is alleged to have said: 'Americans can be relied on to do the right thing...after they have exhausted all other possibilities'. We hope that America and other liberal capitalist states do not wait for an existential crisis to do the right thing. But in the absence of international leadership from the US, it appears that consumers may have to rely on China and other non-liberal capitalist states to sell them the technologies they need to preserve their lifestyles by preventing a dangerous change in the climate.

Notes

1 Economic growth is generally accepted to be the source of social welfare. However, a few leading economists argue that in the rich countries social welfare has become divorced from economic growth. For example, see Jackson (2009), Stiglitz (2002), and Stiglitz et al. (2009).

2   In June 2013 President Obama presented a Climate Action Plan that included regulation of the emissions of power plants. See Chapter 8 for Jeff McGee's cogent summary and White House (2013).
3   A minority of venture capitalists do invest in relatively high risk 'cleantech' ventures. Yet, even they prefer to enter closer to market viability (see Ricadela, 2009).
4   See Mikler and Harrison (2012) on the sectoral contributions to US climate emissions.
5   For DOD complaints about the severity of recent cuts and their probable effect. See http://www.defense.gov/home/features/2013/0213_sequestration/
6   The US Chamber of Commerce, of which many CDP reporters are members, asserts that it is 'standing up for American Enterprise' and vehemently opposes any regulation of economic activity to mitigate climate change or achieve any other environmental or social goal. Indeed, its policies are neoliberal. See http://www.uschamber.com/
7   Giannakouros and Stevis use the term 'liberalism' in Chapter 4, as we do in Chapters 1 and 2, but essentially mean the same as the 'neoliberalism' that McGee defines in Chapter 8.
8   Data available on the EU website at http://ec.europa.eu/clima/policies/g-gas/index_en.htm. National targets are listed at http://ec.europa.eu/clima/policies/g-gas/docs/table_emm_limitation_en.pdf

Bibliography

ACEA (no date), *New Passenger Car Registrations in W. Europe, Breakdown by Specification*, http://www.acea.be/images/uploads/files/20101003_All_Characteristics_1990-201008.pdf, date accessed 18 January 2012.

Adger, W., Lorenzoni, N. and O'Brien, K.L. (2010), *Adapting to Climate Change: Thresholds, Values, Governance* (Cambridge: Cambridge University Press).

Afionis, S. and Bailey, I. (2012), 'Ever Closer Partnerships? European Union Relations with Rapidly Industrializing Countries on Climate Change', in I. Bailey and H. Compston (eds) *Feeling the Heat: The Politics of Climate Policy in Rapidly Industrializing Countries* (Houndmills, Basingstoke: Palgrave Macmillan).

Aguilera, R.V., Williams, C.A., Conley, J.M. and Rupp, D.E. (2006), 'Corporate Governance and Social Responsibility: A Comparative Analysis of the UK and the US', *Corporate Governance – An International Review*, 14(3), 147–158.

Akkermans, D., Castaldi, C. and Los, B. (2009), 'Do "Liberal Market Economies" Really Innovate More Radically than "Coordinated Market Economies"? Hall and Soskice Reconsidered', *Research Policy*, 38, 181–191.

Aldrich, H. and Herker, D. (1977), 'Boundary Spanning Roles and Organization Structure', *Academy of Management Review*, 2(2), 217–230.

Alexander, V. (1996), 'Pictures at an Exhibition', *American Journal of Sociology*, 101(4), 797–839.

Almond, P. (2011), 'The Sub-national Embeddedness of International HRM', *Human Relations*, 64(4), 531–551.

Amable, B. (2003), *The Diversity of Modern Capitalism* (Oxford: Oxford University Press).

Anderson, T. and Leal, D. (2001), *Free Market Environmentalism*, revised edn (Houndmills, Basingstoke: Palgrave Macmillan).

Antia, M., Pantzalis, C. and Park, J.C. (2010), 'CEO Decision Horizon and Firm Performance: An Empirical Investigation', *Journal of Corporate Finance*, 16(3), 288–301.

APEC (2007), *Sydney APEC Leaders Declaration on Climate Change, Energy Security and Clean Development*, http://www.apec.org/Meeting-Papers/Leaders-Declarations/2007/2007_aelm/aelm_climatechange.aspx, date accessed 17 June 2013.

Apollo Alliance (2008), *The New Apollo Program: Clean Energy, Good Jobs: An Economic Strategy for American Prosperity* (San Francisco: Apollo Alliance).

Apollo Alliance (2009), *Make it in America: The Apollo Green Manufacturing Action Plan* (San Francisco: Apollo Alliance).

APP (Asia-Pacific Partnership) (2007), *Asia-Pacific Partnership on Clean Development and Climate: Charter*, http://www.asiapacificpartnership.org/pdf/resources/charter.pdf, date accessed 17 June 2013.

APP (2009), *Flagship Projects*, http://www.asiapacificpartnership.org/pdf/brochure/app-brochure-english.pdf, date accessed 17 June 2013.

APP (2010a), *Asia-Pacific Partnership-Organization*, http://www.asiapacificpartnership.org/english/organization.aspx, date accessed 17 June 2013.
APP (2010b), *Project Roster*, http://www.asiapacificpartnership.org/english/project_roster.aspx, date accessed 17 June 2013.
APP (2010c), *Meetings and Events*, http://www.asiapacificpartnership.org/english/meeting_events.aspx, date accessed 17 June 2013.
Arent, D., Wise, A. and Gelman, R. (2011), 'The Status and Prospects of Renewable Energy for Combating Global Warming', *Energy Economics*, 33, 584–593.
Arthur, W.B. (1989), 'Competing Technologies, Increasing Returns, and Lock-in by Historical Events', *The Economic Journal*, 99(March), 116–131.
Atkinson, R.D. and Hackler, D. (2010), *Economic Doctrines and Approaches to Climate Change Policy* (Washington, DC: Information Technology and Innovation Foundation).
Aulisi, A., Farrell, A., Pershing, J. and VanDeveer, S.D. (2005), *Greenhouse Gas Emissions Trading in US States: Observations and Lessons from the OTC NOx Budget Program* (Washington, DC: World Resources Institute).
Ayres, R.U. and Ayres, L.W. (2002), *A Handbook of Industrial Ecology* (Cheltenham: Edward Elgar).
Bailey, I. (1999), 'Flexibility, Harmonization and the Single Market in EU Environmental Policy', *Journal of Common Market Studies*, 37, 549–571.
Bailey, I., West, J. and Whitehead, I. (2011), 'Out of Sight But Not Out of Mind? Public Perceptions of Wave Energy', *Journal of Environmental Policy and Planning*, 13, 139–157.
Bak, P. (1994), 'Self-Organized Criticality: A Holistic View of Nature', in G. Cowan, D. Pines and D. Meltzer (eds) *Complexity: Metaphors, Models, and Reality* (Reading: Addison-Wesley).
Bak, P. (1996), *How Nature Works: The Science of Self-Organised Criticality* (New York: Copernicus Press).
Bamber, G.J., Lansbury, R. and Wailes, N. (eds) (2011), *International and Comparative Employment Relations: Globalization and Change* (Los Angeles: Sage).
Banig, B. and Trisko, E. (2011), *Low Carbon Jobs Potential*, http://www.greenlaborjournal.org/articles/potential-low-carbon-jobs, date accessed 26 February 2011.
Bansal, P. and Roth, K. (2000), 'Why Companies Go Green: A Model of Ecological Responsiveness', *Academy of Management Journal*, 43(4), 717–736.
Barrett, J., Hoerner, A., Bernow, S. and Dougherty, B. (2002), *Clean Energy and Jobs: A Comprehensive Approach to Climate Change and Energy Policy* (Washington, DC: Economic Policy Institute).
Barry, J. (2007), 'Towards a Model of Green Political Economy: From Ecological Modernization to Economic Security', *International Journal of Green Economics*, 1, 446–464.
Barry, J. and Paterson, M. (2004), 'Globalisation, Ecological Modernization and New Labour', *Political Studies*, 52, 767–784.
Bartlett, C.A. and Ghoshal, S. (1989), *Managing Across Boarders: The Transnational Solution* (Boston: Harvard Business School Press).
Battilana, J., Leca, B. and Boxenbaum, E. (2009), 'How Actors Change Institutions: Towards a Theory of Institutional Entrepreneurship', *The Academy of Management Annals*, 3(1), 65–107.

Baumgartner, F. and Jones, B.D. (1993), *Agendas and Instability in American Politics* (Chicago: University of Chicago Press).

Beach, W., Campbell, K., Kreutzer, D. and Lieberman, B. (2009), *The Economic Impact of Waxman-Markey* (Washington, DC: Heritage Foundation).

Beckert, J. (1999), 'Agency, Entrepreneurs, and Institutional Change: The Role of Strategic Choice and Institutionalized Practices in Organizations', *Organization Studies*, 20(5), 777–799.

Beeson, M. (2009), 'Trading Places? China, the United States and the Evolution of the International Political Economy', *Review of International Political Economy*, 16(4), 729–741.

Benedick, R.E. (1991), *Ozone Diplomacy* (Cambridge: Harvard University Press).

Bergman, N., Connor, P., Markusson, N., Middlemiss, L. and Ricci, M. (2010), 'Bottom-up, Social Innovation for Addressing Climate Change'. Presented at the Sussex Energy Group Conference, www.eci.ox.ac.uk/research/energy/downloads/Bergman%20et%20al%20Social%20Innovation%20WP.pdf, date accessed 10 July 2013.

Bernal, J.D. (1989), 'After Twenty-Five Years: A Reprint of the 1964 Paper by J D Bernal', *Science and Public Policy*, 16(3), 143–151.

Bernstein, S. (2002), *The Compromise of Liberal Environmentalism* (Columbia University Press).

Berwyn, B. (2012), 'Colorado: Governor Hickenlooper Says State Drilling Rules Must Stand, Rejects Call to Withdraw Lawsuit Against Longmont', *Summit County Citizens Voice*, 28 September, http://summitcountyvoice.com/2012/09/28/colorado-gov-hickenlooper-says-state-drilling-rules-must-stand-rejects-call-to-withdraw-lawsuit-against-longmont/, date accessed 10 July 2013.

Block, F. (2008a), 'Crisis and Renewal: The Outlines of a 21st Century New Deal', *Socio-Economic Review*, 9(1), 31–57.

Block, F. (2008b), 'Swimming Against the Current: The Rise of a Hidden Developmental State in the United States', *Politics and Society*, 36(2), 169–206.

Block, F. (2011), 'Innovation and the Invisible Hand of Government', in F. Block and M. Keller (eds) *State of Innovation: The US Government's Role in Technology Development* (Boulder, CO: Paradigm Publishers), pp. 96–108.

Block, F. and Keller, M.R. (2009), 'Where do Innovations Come From? Transformations in the US Economy, 1970–2006', *Socio-economic Review*, 7(3), 459–483.

Block, F. and Keller, M.R. (2011), *State of Innovation: The US Government's Role in Technology Development* (Boulder: Paradigm Publishers).

BlueGreen Alliance (2012), *BlueGreen Alliance Policy Brief: 21st Century Energy in America* (Minneapolis: BlueGreen Alliance).

BlueGreen Alliance (2013), *Jobs21! Good Jobs for the 21st Century*, http://www.bluegreenalliance.org/news/publications/image/Platform-vFINAL.pdf, date accessed 26 November 2013.

BlueGreen Alliance (no date), *BlueGreen Alliance Policy Brief: Clean Water, Good Jobs: BlueGreen Alliance Joint Policy on Water Issues* (Minneapolis: BlueGreen Alliance).

BlueGreen Alliance and American Council for Energy-Efficient Economy (2012), *Gearing Up: Smart Standards Create Good Jobs Building Cleaner Cars* (Minneapolis: BlueGreen Alliance).

BlueGreen Alliance and Apollo Alliance (2011a), *The Michigan Green Manufacturing Action Plan* (Minneapolis: BlueGreen Alliance).
BlueGreen Alliance and Apollo Alliance (2011b), *The Ohio Green Manufacturing Action Plan* (Minneapolis: BlueGreen Alliance).
BlueGreen Alliance and Apollo Alliance (2012), *The California Green Manufacturing Action Plan* (Minneapolis: BlueGreen Alliance).
BlueGreen Alliance, Communications Workers of America, Sierra Club, Progressive States Network (2010), *Networking the Green Economy: How Broadband and Related Technologies can Build a Green Economic Future* (Minneapolis: BlueGreen Alliance), http://www.bluegreenalliance.org/admin/publications/files/Networking-the-Green-Economy.pdf, date accessed 24 November 2013.
Bodansky, D. (2001), 'The History of the Global Climate Change Regime', in U. Luterbacher and D. Sprinz (eds) *International Relations and Global Climate Change* (Cambridge: MIT Press).
Bodansky, D. (2010), 'The Copenhagen Climate Change Conference: A Postmortem', *American Journal of International Law*, 104, 230–240.
Bonsall, T.E. (2002) *Disaster in Dearborn: The Story of the Edsel* (Stanford: Stanford General Books).
Boxenbaum, E. (2006), 'Lost in Translation: The Making of Danish Diversity Management', *American Behavioral Scientist*, 49(7), 939–948.
Boxenbaum, E. and Battilana, J. (2005), 'Importation as Innovation: Transposing Managerial Practices Across Fields', *Strategic Organization*, 3(4), 355–383.
Bradshaw, T.K. and Blakely, E.J. (1999), 'What are Third Wave State Economic Development Efforts? From Incentives to Industrial Policy', *Economic Development Quarterly*, 13(3), 229–244.
Brecher, J. (2012), *Jobs Beyond Coal: A Manual for Communities, Workers and Environmentalists* (Labor Network for Sustainability).
Brenner, R. (2002), *The Boom and the Bubble: The US in the World Economy* (London: Verso).
Breslin, S. (2012), 'Government-industry Relations in China: A Review of the Art of the State', in A. Walter and X. Zhang (eds) *East Asian Capitalism: Diversity, Continuity and Change* (Oxford: Oxford University Press).
Buckley, P.J. (2011), 'International Integration and Coordination in the Global Factory', *Management International Review*, 51(2), 269–283.
Bureau of Labor Statistics (2013), *Economic News Release: Union Members Summary* (United States Department of Labor).
Burgelman, R.A. and Grove, A.S. (2007), 'Let Chaos Reign, then Rein In Chaos – Repeatedly: Managing Strategic Dynamics for Corporate Longevity', *Strategic Management Journal*, 28(10), 965–979.
Burtraw, D. and Woerman, M. (2012), US Status on Climate Change Mitigation, Discussion Paper 12-48 (Washington, DC: Resources for the Future).
Buscher, B. (2010), 'Seeking Telos in the Transfrontier? Neoliberalism and the Transcending of Community Conservation in Southern Africa', *Environment and Planning A*, 42, 644–660.
Bush, V. (1945), *Science, the Endless Frontier: A Report to the President on a Program for Postwar Scientific Research*, 184 (Washington, DC: United States Government Printing Office).

Buttel, F.H. (2000), 'Ecological Modernization as Social Theory', *Geoforum*, 31, 57–65.

Byrne, J., Hughes, K., Rickerson, W. and Kurdgelashvili, L. (2007), 'American Policy Conflict in the Greenhouse: Divergent Trends in Federal, Regional, State, and Local Green Energy and Climate Change Policy', *Energy Policy*, 35(9), 4555–4573.

Cantwell, J., Dunning, J.H. and Lundan, S.L. (2010), 'An Evolutionary Approach to Understanding International Business Activity: The Co-evolution of MNCs and the Institutional Environment', *Journal of International Business Studies*, 41, 567–586.

Cardwell, D. (2013), 'Cities Weighing Taking Over from Private Utilities', *New York Times*, 13 March, http://www.nytimes.com/2013/03/14/business/energy-environment/cities-weigh-taking-electricity-business-from-private-utilities.html, date accessed 10 July 2013.

Carley, S., Brown, A. and Lawrence, S. (2012), 'Economic Development and Energy: From Fad to a Sustainable Discipline', *Economic Development Quarterly*, 26(2), 111–123.

Carley, S., Lawrence, S., Brown, A., Nourafshan, A. and Benami, E. (2010), 'Energy-based Economic Development', *Renewable and Sustainable Energy Reviews*, 15, 282–295.

Carneiro, G. (2013), 'Evaluation of Marine Spatial Planning', *Marine Policy*, 37, 214–229.

Carter, K.E. (2013), 'Capital Structure, Earnings Management, and Sarbanes-Oxley: Evidence from Canadian and U.S. Firms', *Accounting Horizons*, 27(2), 301–318.

Castree, N. (2006), 'From Neoliberalism to Neoliberalisation: Confusions, Consolations and Necessary Illusions', *Environment and Planning A*, 38, 1–6.

CDP (2010), *S&P 100 Carbon Chasm: Analysis based on Carbon Disclosure Project 2007–2009 Responses Drawn from the USA's 100 Largest Companies* (New York: CDP).

CDP (2012), *Business Resilience in an Uncertain, Resource-Constrained World: CDP Global 500 Climate Change Report 2012* (New York: CDP).

CDP (2012a), *CDP Global 500 Climate Change Report 2012: Business Resilience in an Uncertain, Resource-constrained World* (London: CDP).

CDP (2012b), *CDP S&P 500 Climate Change Report 2012: Accelerating Progress Toward a Lower-carbon Future* (London: CDP).

CDP (2012c), *CDP Asia ex-Japan Climate Change Report 2012: Increased Opportunities from Emerging Regulation* (London: CDP).

Center for Climate and Energy Solutions (2012), 'Climate Debate in Congress', August, http://www.c2es.org/federal/congress, date accessed 10 July 2013.

Cerny, P. (1997), 'Paradoxes of the Competition State: The Dynamics of Political Globalization', *Government and Opposition*, 32(2), 251–274.

Chang, H. (2002), *Kicking Away the Ladder* (London: Anthem).

Chang, H. (2003), *Globalisation, Economic Development and the Role of the State* (London and New York: Zed Books).

Chen, G., Firth, M. and Xu, L. (2009), 'Does the Type of Ownership Control Matter? Evidence from China's Listed Companies', *Journal of Banking and Finance*, 33(1), 171–181.

Christoff, P. (1996), 'Ecological Modernization, Ecological Modernities', *Environmental Politics*, 5(3), 476–500.

Christoff, P. (2013), 'Climate Discourse Complexes, National Climate Regimes and Australian Climate Policy', *Australian Journal of Politics and History*, forthcoming.

Ciftci, M., Lev, B. and Radhakrishnan, S. (2011), 'Is Research and Development Mispriced or Properly Risk Adjusted?', *Journal of Accounting, Auditing and Finance*, 26(1), 81–116.

Clapp, J. (1998), 'The Privatization of Global Environmental Governance: ISO 14000 and the Developing World', *Global Governance*, 4, 295–316.

Clarke, S.E. and Gaile, G.L. (1992), 'The Next Wave: Postfederal Local Economic Development Strategies', *Economic Development Quarterly*, 6, 187–198.

Climate Change Technology Program (CCTP) (2006), *Strategic Plan*, http://www.climatetechnology.gov.

Cobb, J.C. and Stueck, W. (eds) (2005), *Globalization and the American South* (Athens, Georgia: University of Georgia Press).

Cogan, D.G. (2006), *Corporate Governance and Climate Change: Making the Connection* (Boston: CERES).

Cohen, M.J. (2006), 'Ecological Modernization and its Discontents: The American Environmental Movement's Resistance to an Innovation-driven Future', *Futures*, 38, 528–547.

Cohen-Rosenthal, E. (1997), 'Sociotechnical Systems and Unions: Nicety or Necessity?', *Human Relations*, 50(5), 585–604.

Cohen-Rosenthal, E., Fabens, B. and McGalliard, T. (1998), 'Labor and Climate Change: Dilemmas and Solution', *New Solutions*, 8(3), 343–363.

Committee on Climate Change (2012), *Renewable Energy Review*, http://www.publications.parliament.uk/pa/cm201012/cmselect/cmenergy/1624/1624.pdf, date accessed 12 April 2013.

Committee on Climate Change Science and Technology Integration (CCCSTI) (2009), 'Strategies for the Commercialization and Deployment of Greenhouse Gas Intensity-Reducing Technologies and Practices'.

Contractor, F. (2012), *7 Reasons to Expect US Manufacturing Resurgence* (YaleGlobal Online).

Cooke, P. (2010), 'Socio-technical Transitions and Varieties of Capitalism: Green Regional Innovation and Distinctive Market Niches', *J Knowl Econ*, 1, 239–267.

Crouch, C. (2005a), 'Models of Capitalism', *New Political Economy*, 10(4), 339–456.

Crouch, C. (2005b), *Capitalist Diversity and Change: Recombinant Governance and Institutional Entrepreneurs* (Oxford: Oxford University Press).

Crouch, C. and Streeck, W. (eds) (1997), *Political Economy of Modern Capitalism: Mapping Convergence and Diversity* (London: Sage).

Crowley, K. (2013), 'Irresistible Force? Achieving Carbon Pricing in Australia', *Australian Journal of Politics and History*, forthcoming.

Cushman, J.H.J. (1998), 'New Policy Center Seeks to Steer the Debate on Climate Change', *The New York Times*, 8 May.

D'Aunno, T., Sutton, R.I. and Price, R.H. (1991), 'Isomorphism and External Support in Conflicting Institutional Environments: A Study of Drug Abuse Treatment Units', *Academy of Management Journal*, 34(3), 636–661.

Darnall, N., Henriques, I. and Sadorsky, P. (2010), 'Adopting Proactive Environmental Strategy: The Influence of Stakeholders and Firm Size', *Journal of Management Studies*, 47(6), 1072–1094.

David, P.A. (1985), 'Clio and the Economics of Qwerty', *The American Economic Review*, 75(2), 332–337.

De Sombre, E. (2011), 'The United States and Global Environmental Politics: Domestic Sources of US Unilateralism', in R. Axelrod, S. Vandeveer and D. Downie (eds), *The Global Environment: Institutions, Law and Policy*, 3rd edn (Washington, DC: CQ Press).

Debbage, K.G. and Jacob, F. (2011), 'Renewable Energy in North Carolina: The Potential Supply Chain Connections to Existing Renewable Energy Efficiency Firms', *Southeastern Geographer*, 51(1), 69–88.

DeGennaro, R., Dwyer, P. and Gerald, P. (2010), *Expected Returns to Stock Investments by Angel Investors in Groups*, http://ssrn.com/abstract=1360817 or http://dx.doi.org/10.2139/ssrn.1360817.

Dejean, F., Gond, J. and Leca, B. (2004), 'Measuring the Unmeasured: An Institutional Entrepreneur Strategy in an Emerging Industry', *Human Relations*, 57(6), 741–746.

Delbridge, R. and Edwards, T. (2008), 'Challenging Conventions: Roles and Processes During Non-isomorphic Institutional Change', *Human Relations*, 61(3), 299–325.

Delmas, M.A. and Toffel, M.W. (2008), 'Organizational Responses to Environmental Demands: Opening the Black Box', *Strategic Management Journal*, 29(10), 1027–1055.

Deng, P. (2009), 'Why do Chinese Firms Tend to Acquire Strategic Assets in International Expansion?', *Journal of World Business*, 44(1), 74–84.

Department for Communities and Local Government (2012), *National Planning Policy Framework 2012* (London: Stationery Office).

Department of Energy (2012a), 'Funding Opportunities', *Energy.Gov*, http://energy.gov/public-services/funding-opportunities, date accessed 10 July 2013.

Department of Energy (2012b), *Loans Program Office*, https://lpo.energy.gov/?page_id=45, date accessed 10 July 2013.

Department of Energy and Climate Change (2011), *Consultation on Proposals for the Levels of Banded Support Under the Renewables Obligation for the Period 2013–17 and the Renewables Obligation Order 2012* (London: Stationery Office).

Depledge, J. (2005), 'Against the Grain: United States and the Climate Change Regime', *Global Change Peace and Security*, 17(1), 11–27.

Deutsche Bank (2004), *The Drivers: How to Navigate the Auto Industry* (Frankfurt am Main: Deutsche Bank AG).

Deutz, P. (2009), 'Producer Responsibility in a Sustainable Development Context: Ecological Modernisation or Industrial Ecology', *The Geographical Journal*, 175(4), 274–285.

DiMaggio, P.J. and Powell, W.W. (1983), 'The Iron Cage Revisited: Institutional Isomorphism and Collective Rationality in Organizational Fields', *American Sociological Review*, 48(2), 147–160.

Dore, R. (2000), 'Will Global Capitalism be Anglo-Saxon Capitalism?', *New Left Review*, 6(November/December), 101–119.

Dore, R., Lazonick, W. and O'Sullivan, M. (1999), 'Varieties of Capitalism in the Twentieth Century', *Oxford Review of Economic Policy*, 15(4), 102–120.

Dosi, G. (1982), 'Technological Paradigms and Technological Trajectories: A Suggested Interpretation of the Determinants and Directions of Technical Change', *Research Policy*, 11(3), 147–162.

Dowell, G., Hart, S. and Yeung, B. (2000), Do Corporate Global Environmental Standards Create or Destroy Market Value?', *Management Science*, 46, 1059–1074.

Downer, A. (2005), *Minister for Foreign Affairs; Press Conference, ITECC*, Vientiane Laos, http://pandora.nla.gov.au/pan/25167/20070502-0000/www.foreignminister.gov.au/transcripts/2005/050728_itecc.html, date accessed 17 June 2013.

Drezner, D.W. (2007), *All Politics is Global: Explaining International Regulatory Regimes* (Princeton: Princeton University Press).

Driesen, D. (ed.) (2009), *Economic Thought and US Climate Change Policy* (MIT Press).

Driesen, D. (ed.) (2010), *Economic Thought and US Climate Change Policy* (Cambridge, MA: MIT Press).

Dryzek, J. (2005), *Politics of the Earth; Environmental Discourses* (Oxford: Oxford University Press).

Dryzek, J.S. (1997), *The Politics of the Earth: Environmental Discourses* (Oxford: Oxford University Press).

Dryzek, J.S., Downes, D., Hunold, C., Schlosberg, D. and Hernes, H. (2003), *Green States and Social Movements: Environmentalism in the United States, United Kingdom, Germany, and Norway* (Oxford: Oxford University Press).

Dunning, J.H., van Hoesel, R. and Narula, R. (1998), 'Third World Multinationals Revisited: New Developments and Theoretical Implications', in J.H. Dunning (ed.) *Globalization, Trade and Foreign Direct Investment* (Oxford: Pergamon).

Dye, L. (2010), 'Global Warming and the Pollsters: Who's Right?', *ABC News*, 16 June, http://abcnews.go.com/Technology/DyeHard/global-warming-polls-climate-change/story?id=10921583, date accessed 5 May 2012.

Edwards, P.N. and Lahsen, M.H. (1999), *Climate Science and Politics in the United States* (Michigan: School of Information University of Michigan).

Ehrlich, P.R. (2006), 'Environmental Science Input to Public Policy', *Social Research*, 73(3), 915–948.

Eisinger, P. (1990), 'Do the American States do Industrial Policy?', *British Journal of Political Science*, 20(4), 509–535.

Eisinger, P. (1995), 'State Economic Development in the 1990's: Politics and Policy Learning', *Economic Development Quarterly*, 9(2), 146–158.

Eldredge, N. and Gould, S.J. (1972), 'Punctuated Equilibria: An Alternative to Phyletic Gradualism', in T.J.M. Schopf (ed.) *Models in Paleobiology* (San Francisco: Freeman Cooper).

eROC (2011), *ROC Track Record*, http://www.e-roc.co.uk/trackrecord.htm, date accessed 23 November 2012.

Esping-Andersen, G. (1990), *The Three Worlds of Welfare Capitalism* (Cambridge: Polity Press).

Etzkowitz, H. (2003), 'Innovation in Innovation: The Triple Helix of University-Industry-Government Relations', *Social Science Information*, 42, 293–337.

European Commission (2013), *Report from the Commission to the European Parliament, the Council, the European Economic and Social Committee and the Committee of the Regions: Renewable Energy Progress Report* (Brussels: European Commission).

European Wind Energy Association (2011), *EU Energy Policy to 2050: Achieving 80–95% Emissions Reductions* (Brussels: European Wind Energy Association).

Executive Office of the President (2013), *The President's Climate Action Plan* (Washington, DC: White House).

Fennell, M.L. and Alexander, J.A. (1987), 'Organizational Boundary Spanning in Institutionalized Environments', *Academy of Management Journal*, 30(3), 456–476.

Festinger, L. (1957), *A Theory of Cognitive Dissonance* (Evanston: Row, Peterson).

Finet, D. (1993), 'Effects of Boundary Spanning Communication on the Sociopolitical Delegitimizing of an Organization', *Management Communication Quarterly*, 7(1), 31–36.

Fletcher, S., McKinley, E., Buchan, K., Smith, N. and McHugh, K. (2013), 'Effective Practice in Marine Spatial Planning: A Participatory Evaluation of Experience in Southern England', *Marine Policy*, 39, 341–348.

Fligstein, N. and Zhang, J. (2011), 'A New Agenda for Research on the Trajectory of Chinese Capitalism', *Management and Organization Review*, 7, 39–62.

Forsgren, M., Holm, U. and Johanson, J. (2005), *Managing the Embedded Multinational: A Business Network View* (Cheltenham: Edward Elgar).

Fortune (2012), *Global 500 2012*, http://money.cnn.com/magazines/fortune/global500/, date accessed 2 July 2013.

Foster, D. (2010), 'BlueGreen Alliance: Building a Coalition for a Green Future in the United States', *International Journal of Labour Research*, 2(2), 233–244.

Franz, W. (1998), *Science, Skeptics, and Non-State Actors in the Greenhouse* (Cambridge: Harvard University).

Freeman, C. (1979), 'The Determinants of Innovation: Market Demand, Technology, and the Response to Social Problems', *Futures*, 11, 206–215.

Freeman, C. (1987), *Technology and Economic Performance: Lessons from Japan* (London: Pinter).

Freeman, C. (1992), *The Economics of Hope: Essays on Technical Change, Economic Growth, and the Environment* (London: Pinter Publishers).

Freeman, C. (1995), 'The National System of Innovation in Historical Perspective', *Cambridge Journal of Economics*, 19, 5–24.

Freeman, R.E. (2010[1984]), *Strategic Management: A Stakeholder Approach* (Cambridge: Cambridge University Press).

Friedman, M. (1970), 'The Social Responsibility of Business is to Increase Profits', *New York Times Magazine*, 13 September.

Frondel, M., Ritter, N., Schmidt, C.M. and Vance, C. (2009), *Economic Impacts from the Promotion of Renewable Energy Technologies: The German Experience*, Ruhr Economic Papers #156, Ruhr-Universität Bochum, Germany, November, available at: http://repec.rwi-essen.de/files/REP_09_156.pdf, accessed 12 May 2012.

Frosch, R.A. and Gallopoulos, N. (1989), 'Strategies for Manufacturing', *Scientific American*, 261(3), 144–152.

Fuchs, E. (2010), 'Rethinking the Role of the State in Technology Development: DARPA and the Case for Embedded Network Governance', *Research Policy*, 39(9), 1133–1147.

Furukawa, R. and Goto, A. (2006), 'The Role of Corporate Scientists in Innovation', *Research Policy*, 35(1), 24–36.

Garnaut, R. (2008), *The Garnaut Climate Change Review: Final Report* (Melbourne: Cambridge University Press).

Gault, F. and Huttner, S. (2008), 'A Cat's Cradle for Policy', *Nature*, 455, 462–463.

Geels, F.W. (2004), 'From Sectoral Systems of Innovation to Socio-technical Systems: Insights About Dynamics and Change from Sociology and Institutional Theory', *Research Policy*, 33(6), 897–920.

Gelbspan, R. (1997), *The Heat is On* (Reading: Addison Wesley).

Gereffi, G., Dubay, K. and Lowe, M. (2008), *Manufacturing Climate Solutions: Carbon-Reducing Technologies and U.S. Jobs* (Durham: Center on Globalization, Governance and Competitiveness, Duke University).

Gertner, J. (2011), 'Does America Need Manufacturing?', *New York Times*, 24 August.

Gibbs, D. (2000), 'Ecological Modernization, Regional Economic Development and Regional Development Agencies', *Geoforum*, 31(1), 9–19.

Giddens, A. (2011), *The Politics of Climate Change*, Second Edition (Cambridge: Polity Press).

Gieryn, T. (1999), *Cultural Boundaries of Science: Credibility on the Line* (Chicago: University Chicago Press).

Gilson, R.J. (1996), 'Corporate Governance and Economic Efficiency: When do Institutions Matter?', *Washington University Law Quarterly*, 74, 327–345.

Gilson, R.J. (2006), 'Controlling Shareholders and Corporate Governance: Complicating the Comparative Taxonomy', *Harvard Law Review*, 119(6), 1641–1679.

Goodin, R.E. (2003), 'Choose Your Capitalism?', *Comparative European Politics*, 1(2), 203–213.

Gouldson, A. and Murphy, J. (1997), 'Ecological Modernization: Restructuring Industrial Economies', in M. Jacobs (ed.) *Greening the Millennium? The New Politics of the Environment* (Oxford: Blackwell Publishers).

Government of Republic of South Korea (2010), *Asia-Pacific Partnership on Clean Development and Climate*, http://app-korea.kemco.or.kr/, date accessed 17 June 2013.

Graham, D. and Glaister, S. (2002), 'The Demand for Automobile Fuel: A Survey of Elasticities', *Journal of Transport Economics and Policy*, 36(1), 1–26.

Graham, D. and Glaister, S. (2004), 'Road Traffic Demand Elasticity Estimates: A Review', *Transport Reviews*, 24(3), 261–274.

Greening, L.A., Greene, D.L. and Difiglio, C. (2000), 'Energy Efficiency and Consumption – The Rebound Effect – A Survey', *Energy Policy*, 28(6–7), 389–401.

Greenwood, R. and Suddaby, R. (2006), 'Institutional Entrepreneurship in Mature Fields: The Big Five Accounting Firms', *Academy of Management Journal*, 49(1), 27–48.

Greenwood, R., Suddaby, R. and Hinings, C.R. (2002), 'Theorizing Change: The Role of Professional Associations in the Transformation of Institutionalized Fields', *Academy of Management Journal*, 45(1), 58–80.

Grossman, R. and Daneker, G. (1979), *Energy, Jobs and the Economy* (Boston: Alyson Publishers, Inc.)

Grubb, M. (2005), 'Technology Innovation and Climate Change Policy: An Overview of Issues and Options', *Keio Economic Studies*, XLI(2), 103–132.

Guber, D. (2003), *Grassroots of a Green Revolution: Polling America on the Environment* (Cambridge: MIT Press).

Haas, P.M. (1992), 'Banning Chlorofluorocarbons: Epistemic Community Efforts to Protect Stratospheric Ozone', *International Organization*, 46(1), 211–235.

Haas, R., Resch, G., Panzer, C., Busch, S., Ragwitz, M. and Held, A. (2011), 'Efficiency and Effectiveness of Promotion Systems for Electricity Generation from Renewable Energy Sources: Lessons from EU Countries', *Energy*, 36(2), 186–182, 193.

Hajer, M.A. (1995), *The Politics of Environmental Discourse: Ecological Modernization and the Policy Process* (New York: Oxford University Press).

Hall, P. and Soskice, D. (2001a), 'An Introduction to Varieties of Capitalism', in P. Hall and D. Soskice (eds) *Varieties of Capitalism: The Institutional Foundations of Comparative Advantage* (Oxford: Oxford University Press).

Hall, P. and Soskice, D. (eds) (2001b), *Varieties of Capitalism: The Institutional Foundations of Comparative Advantage* (Oxford: Oxford University Press).

Hall, P.A. and Thelen, K. (2009), 'Institutional Change in Varieties of Capitalism', *Socio-Economic Review*, 7(1), 7–34.

Halper, S. (2010), *The Beijing Consensus: How China's Authoritarian Model Will Dominate the Twenty-First Century* (New York: Basic Books).

Hamilton, C. (2007), *Scorcher: The Dirty Politics of Climate Change* (Melbourne: Black Inc. Agenda).

Hampden-Turner, C. and Trompenaars, A. (1993), *The Seven Cultures of Capitalism: Value Systems for Creating Wealth in the United States, Japan, Germany, France, Britain, Sweden and the Netherlands* (New York: Currency Doubleday).

Hancke, B. (2009), *Debating Varieties of Capitalism: A Reader* (Oxford: Oxford University Press).

Handwerker, E.W., Kim, M.M. and Mason, L.G. (2011), 'Domestic Employment in U.S.-based Multinational Companies', *Monthly Labor Review Online*, 134(10), 3–15.

Hansen, S.B. (1989), 'Industrial Policy and Corporatism in the American States', *Governance*, 2(2), 172–197.

Harrington, W. and McConnell, V. (2003), *Motor Vehicles and the Environment* (Washington, DC: Resources for the Future), http://www.rff.org/Documents/RFF-RPT-carsenviron.pdf, date accessed 29 June 2013.

Harris, P. (2009), 'Beyond Bush: Environmental Politics and Prospects for US Climate Policy', *Energy Policy*, 37(3), 966–971.

Harrison, N.E. (2014), *Sustainable Capitalism and the Pursuit of Well-being* (London: Routledge).

Hart, D.M. (2008), 'The Politics of "Entrepreneurial" Economic Development Policy of States in the US', *Review of Policy Research*, 25(2), 149–168.

Hartman, T. (2011), *A Blueprint for a New Energy Economy*, http://rechargecolorado.com/images/uploads/pdfs/Ritter_Energy_Book_F_ForWeb.pdf, date accessed 11 April 2011.

Hassol, S.J. (2011), *Emissions Reductions Needed to Stabilize Climate*, Statement for the Presidential Climate Action Project, http://climatecommunication.org/wp-content/uploads/2011/08/presidentialaction.pdf, date accessed 5 May 2012.

Haufler, V.A. (1998), *Policy Trade-offs and Industry Choice: Hedging Your Bets on Global Climate Change* (Boston: American Political Science Association Annual Meeting).

Hawken, P. and Lovins, A.B. (1999), *Natural Capitalism: Creating the Next Industrial Revolution* (Boston: Little, Brown and Co).

Hayden, G.F., Douglas, K.C. and Williams, S.C. (1985), 'Industrial Policy at the State Level in the United States', *Journal of Economic Issues*, 19(2), 383–396.

Hayek, F. (1976), *Law, Legislation and Liberty, Vol. 2: The Mirage of Social Justice* (London: Routledge).

Hayek, F.A. (1994[1944]), *The Road to Serfdom* (Chicago: The University of Chicago Press).

Heidenreich, M. (2012), 'The Social Embeddedness of Multinational Companies: A Literature Review', *Socio-Economic Review*, 10, 549–579.

Helm, D. (2012), *The Carbon Crunch: How We're Getting Climate Change Wrong – and How to Fix It* (New Haven: Yale University Press).

Heynen, N., McCarthy, J., Prudham, S. and Robbins, P. (eds) (2007), *Neoliberal Environments: False Promises and Unnatural Consequences* (London: Routledge).

Hirukawa, M. and Ueda, M. (2011), 'Venture Capital and Innovation: Which is First?', *Pacific Economic Review*, 16(4), 421–465.

HM Government (2012), *Draft Energy Bill: May 2012* (London: Stationery Office).

Hobsbawm, E. (1999), *Industry and Empire: From 1750 to the Present Day* (New York: New Press).

Hoffman, A.J. (1997), *From Heresy to Dogma: An Institutional History of Corporate Environmentalism* (San Francisco: The New Lexington Press).

Hofstede, G. (2003), *Culture's Consequences, Comparing Values, Behaviors, Institutions, and Organizations across Nations*, 2nd edn (Thousand Oaks: Sage Publications).

Hogler, R. (2004), *Employment Relations in the United States: Law, Policy and Practice* (London: Sage).

Hollingsworth, J. (1997), 'Continuities and Changes in Social Systems of Production: The Cases of Japan, Germany, and the United States', in J. Hollingsworth and R. Boyer (eds) *Contemporary Capitalism: The Embeddedness of Institutions* (Cambridge: Cambridge University Press).

Hollingsworth, J.R. and Boyer, R. (eds) (1997), *Contemporary Capitalism: The Embeddedness of Institutions* (Cambridge: Cambridge University Press).

Holm, P. (1995), 'The Dynamics of Institutionalization: Transformation Processes in Norwegian Fisheries', *Administrative Science Quarterly*, 40, 398–422.

Howell, C. (2003), 'Varieties of Capitalism: And Then There was One?', *Comparative Politics*, 36(1), 103–124.

Hrynyshyn, D. and Ross, S. (2010), 'Canadian Autoworkers, the Climate Crisis, and the Contradictions of Social Unionism', *Labor Studies Journal*, 36(1), 5–36.

Huber, J. (2000), 'Towards Industrial Ecology: Sustainable Development as a Concept of Ecological Modernization', *Journal of Environmental and Policy Planning*, 2, 269–285.

Hunold, C. and Dryzek, J.S. (2005), 'Green Political Strategy and the State: Combining Political Theory and Comparative History', in J. Barry and R. Eckersley (eds) *The State and the Global Ecological Crisis* (Boston: MIT Press).

Hurt, S. (2010), 'The Military's Hidden Hand: Examining the Dual-Use Origins of Agricultural Biotechnology in the American Context, 1969–1972', in F. Block and M. Keller (eds) *State of Innovation: The US Government's Role in Technology Development* (Boulder: Paradigm).

IEA (2009), *Energy Prices and Taxes: Quarterly Statistics, Second Quarter 2009* (Paris: OECD/IEA).

Jackson, G. (2010), *Understanding Corporate Governance in the United States: An Historical and Theoretical Reassessment* (Düsseldorf: Hans Böckler Stiftung).

Jackson, G. and Apostolakou, A. (2010), 'Corporate Social Responsibility in Western Europe: An Institutional Mirror or Substitute?', *Journal of Business Ethics*, 94, 371–394.

Jackson, G. and Deeg, R. (2008), 'Comparing Capitalisms: Understanding Institutional Diversity and its Implications for International Business', *Journal of International Business Studies*, 39, 540–561.

Jackson, G. and Petraki, A. (2011), *Understanding Short-Termism: The Role of Corporate Governance* (Stockholm: Glasshouse Forum).

Jackson, T. (2009), *Prosperity Without Growth: Economics for a Finite Planet* (London: Earthscan).

JAMA (2011), *Motor Vehicle Statistics of Japan 2011* (Tokyo: JAMA), http://www.jama-english.jp/publications/MIJ2011.pdf, date accessed 19 January 2012.

Jänicke, M. (1991), *The Political System's Capacity for Environmental Policy* (Berlin: Department of Environmental Politics, Free University Berlin).

Jänicke, M. and Lindemann, S. (2010), 'Governing Environmental Innovations', *Environmental Politics*, 19(1), 127–141.

Jasanoff, S. (1990), *The Fifth Branch* (Cambridge: Harvard University Press).

Jasanoff, S. (1991), 'Cross-national Differences in Policy Implementation', *Evaluation Review*, 15(1), 103–119.

Jha, A. (2010), 'Global Emissions of Carbon Dioxide Drop 1.3%, Say International Scientists', *The Guardian*, 22 November, http://www.guardian.co.uk/environment/2010/nov/21/carbon-emissions-fall-report, date accessed 17 June 2013.

Jiusto, S. (2010), 'Spatial Indeterminacy and the Construction of Environmental Knowledge', *The Geographical Journal*, 176(3), 214–226.

Johnson, C.A. (1995), *Japan: Who Governs? The Rise of the Developmental State* (New York: Norton).

Johnston, J. (1997), *Driving America: Your Car, Your Government, Your Choice* (Washington, DC: AEI Press).

Jordan, A. and Adelle, C. (2012), 'EU Environmental Policy: Contexts, Actors and Policy Dynamics', in A. Jordan and C. Adelle (eds) *Environmental Policy in the European Union: Actors, Institutions and Processes* (Abingdon: Routledge).

Jordan, A., Huitema, D. and van Asselt, H. (2010), 'Climate Change Policy in the European Union: An Introduction', in A. Jordan, D. Huitema, H. van Asselt, T. Rayner and F. Berkhout (eds) *Climate Change Policy in the European Union: Confronting the Dilemmas of Mitigation and Adaptation* (Cambridge: Cambridge University Press).

Jordan, A., Wurzel, R. and Zito, A. (eds) (2003) *'New' Instruments of Environmental Governance? National Experiences and Prospects* (London: Frank Cass).

Jorgensen, L. (2012) 'Gov Pitches Increase for Economic Development Incentives', *The Colorado Observer*, 12 December, http://thecoloradoobserver.com/2012/12/gov-pitches-increase-for-economic-development-incentives/, date accessed 12 December 2012.

Kablan, Y. and Michalak, S. (2012), *National Policy Framework for Marine Renewable Energy within France* (Brest: Parc Naturel Marin d'Iroise).

Kang, N. and Moon, J. (2012), 'Institutional Complementarity between Corporate Governance and Corporate Social Responsibility: A Comparative Institutional Analysis of Three Capitalisms', *Socio-Economic Review*, 10(1), 85–108.

Katz, H. and Colvin, A. (2011), 'Employment Relations in the United States', in G. Bamber, R. Lansbury and N. Wailes (eds) *International and Comparative Employment Relations: Globalization and Change* (Los Angeles: Sage).

Kauffman, S. (1995), *At Home in the Universe: The Search for the Laws of Self-Organization and Complexity* (New York: Oxford University Press).

Keller, J. (2012), 'Military Technology Research and Development Set to Continue Three-Year Decline in 2013', *Military and Aerospace News*, 21 February.

Keller, K.H. (2008), 'From Here to There in Information Technology: The Complexities of Innovation', *American Behavioral Scientist*, 52(1), 97–106.

Keller, M. (2010), 'The CIA's Pioneering Role in Public Venture Capital Initiatives', in F. Block and M. Keller (eds) *State of Innovation: The US Government's Role in Technology Development* (Boulder: Paradigm).

Kellow, A. and Zito, A. (2002), 'Steering through Complexity: EU Environmental Regulation in the International Context', *Political Studies*, 50, 4–60.

Kelly, D. and Amburgey, T.L. (1991), 'Organizational Inertia and Momentum: A Dynamic Model of Strategic Change', *Academy of Management Journal*, 34(3), 591–612.

Kempton, W. and Craig, P.P. (1993), 'European Perspectives on Global Climate Change', *Environment*, 35(3), 17–45.

Kerkhoff, L.V. and Lebel, L. (2006), 'Linking Knowledge and Action for Sustainable Development', *Annual Review of Environment and Resources*, 31, 445–477.

Kern, F. (2011), 'Ideas, Institutions and Interests: Explaining Policy Divergence in Fostering "System Innovations" Towards Sustainability', *Environment and Planning C: Government and Policy*, 29, 1116–1134.

Kilduff, M. (2005), 'The Reproduction of Inertia in Multinational Corporations', in S. Ghoshal and D.E. Westney (eds) *Organization Theory and the Multinational Corporation*, 2nd edition (Basingstoke: Palgrave Macmillan).

King Jr, N. (2009), 'Venture Capital: New VC Force', *The Wall Street Journal*, 15 December, http://online.wsj.com/article/SB126074549073889853.html, accessed 10 July 2013.

Kitzing, L., Mitchell, C. and Morthorst, P.E. (2012), 'Renewable Energy Policies in Europe: Converging or Diverging?', *Energy Policy*, 51, 192–201.

Klyza, C. and Sousa, D. (2008) *American Environmental Policy, 1990–2006: Beyond Gridlock* (Cambridge, MA: MIT Press)

Knittel, C.R. (2011), 'Automobiles on Steroids: Product Attribute Trade-Offs and Technological Progress in the Automobile Sector', *American Economic Review*, 101(7), 3368–3369.

Koch, M. (2012), *Capitalism and Climate Change: Theoretical Discussion, Historical Development and Policy Responses* (Houndmills, Basingstoke: Palgrave Macmillan).

Kochan, T. (1975), 'Determinants of Power of Boundary Units in Organizational Bargaining Relations', *Administrative Science Quarterly*, 2(3), 434–452.

Kogut, B. (2005), 'Learning, or the Importance of Being Inert: Country Imprinting and International Competition', in S. Ghoshal and D.E. Westney (eds) *Organization Theory and the Multinational Corporation*, 2nd edition (Basingstoke: Palgrave Macmillan).

Kogut, B. and Zander, U. (1993), 'Knowledge of the Firm and the Evolutionary Theory of the Multinational Corporation', *Journal of International Business Studies*, 19(3), 625–645.

Kojola, E. (2009), 'Trade Unions and Green Jobs in a post-Fordist Economy: Just Rhetoric or a Fundamental Shift?', *Politics*, Oberlin College, BA, 131.

Kolk, A. and Pinkse, J. (2008), 'A Perspective on Multinational Enterprises and Climate Change: Learning from "An Inconvenient Truth"?', *Journal of International Business Studies*, 39, 1359–1378.

Kolk, A., Levy, D. and Pinkse, J. (2008), 'Corporate Responses in an Emerging Climate Regime: The Institutionalization and Commensuration of Carbon Disclosure', *European Accounting Review*, 17(4), 719–745.

Kruck, C., Borchers, J. and Weingart, P. (1999), *Climate Research and Climate Politics in Germany: Assets and Hazards of Consensus-based Risk Management* (Germany: Department of Sociology, University of Bielefeld).

Krug, B. and Hendrischke, H. (2008), 'Framing China: Transformation and Institutional Change through Co-evolution', *Management and Organization Review*, 4(1), 81–108.

Lam, J., Hills, P. and Welford, R. (2005), 'Ecological Modernisation, Environmental Innovation and Competitiveness: The Case of Public Transport in Hong Kong', *Int. J. Innovation and Sustainable Development*, 1(1/2), 103–126.

Lamertz, K. and Heugens, P.P.M.A.R. (2009), 'Institutional Translation through Spectatorship: Collective Consumption and Editing of Symbolic Organizational Texts by Firms and Their Audiences', *Organization Studies*, 30(11), 1249–1279.

LaMonica, M. (2012), 'Should the Government Support Applied Research?', *Technology Review*, 115(6), 81–82.

Lane, C. and Wood, G. (2009), 'Capitalist Diversity and Diversity within Capitalism', *Economy and Society*, 38(4), 531–551.

Laurie, N. (2005), 'Putting the Messiness Back In: Towards a Geography of Development as Creativity', *Singapore Journal of Tropical Geography*, 26, 32–35.

Lawrence Berkeley Laboratory (2010), 'Technology Transfer: From Berkeley Lab to the Marketplace', Lawrence Berkeley National Laboratory.

Leca, B., Battilana, J. and Boxenbaum, E. (2008), *Agency and Institutions: A Review of Institutional Entrepreneurship*, Working Paper 08-096 (Boston: Harvard Business School).

Leifer, R. and Delbecq, A. (1978), 'Organizational/Environmental Interchange: A Model of Boundary Spanning Activity', *Academy of Management Review*, 3(1), 40–50.

Levin, S.A. (1999), *Fragile Dominion: Complexity and the Commons* (Reading: Perseus Books).
Levy, D. and Rothenberg, S. (2002), 'Heterogeneity and Change in Environmental Strategy: Technological and Political Responses to Climate Change in the Global Automobile Industry', in A. Hoffman and M. Ventresca (eds) *Organizations, Policy and the Natural Environment: Institutional and Strategic Perspectives* (Stanford: Stanford University Press).
Levy, D.L. (1997), 'Environmental Management as Political Sustainability', *Organization and Environment*, 10(2), 126–147.
Levy, D.L. and Egan, D. (1998), 'Capital Contests: National and Transnational Channels of Corporate Influence on the Climate Change Negotiations', *Politics and Society*, 26(3), 337–361.
Levy, D.L. and Kolk, A. (2002), 'Strategic Responses to Global Climate Change: Conflicting Pressures on Multinationals in the Oil Industry', *Business and Politics*, 4(3), 275–300.
Levy, O., Beechler, S., Taylor, S. and Boyacigiller, N.A. (2007) 'What We Talk About When We Talk About "Global Mindset": Managerial Cognition in Multinational Corporations', *Journal of International Business Studies*, 38, 231–258.
Lewis, J.I. and Wiser, R.H. (2007), 'Fostering a Renewable Energy Technology Industry: An International Comparison of Wind Industry Policy Support Mechanisms', *Energy Policy*, 35(3), 1844–1857.
Li, W. and Zhang, R. (2010), 'Corporate Social Responsibility, Ownership Structure and Political Interference: Evidence from China', *Journal of Business Ethics*, 96(4), 631–645.
Li, Y. and Zahra, S.A. (2012) 'Formal Institutions, Culture, and Venture Capital Activity: A Cross-Country Analysis', *Journal of Business Venturing*, 27(1), 95–111.
Lifset, R. and Graedel, T.E. (2002), 'Industrial Ecology: Goals and Definitions', in R.U. Ayres and L.W. Ayres (eds) *A Handbook of Industrial Ecology* (Cheltenham: Edward Elgar).
Ling, Z. (2006), *The Lenovo Affair: The Growth of China's Computer Giant and Its Takeover of IBM-PC* (Singapore: John Wiley and Sons).
Lipietz, A. (1989), *Choisir l'audace: une alternative pour le XXIe siècle* (Paris: Editions La Découverte).
Liverman, D.M. (2009), 'Conventions of Climate Change: Constructions of Danger and the Dispossession of the Atmosphere', *Journal of Historical Geography*, 35, 279–296.
Loughlin, J. (2008), 'The Slow Emergence of the French Regions', *Policy and Politics*, 36, 559–571.
Lounsbury, M., Ventresca, M. and Hirsch, P.M. (2003), 'Social Movements, Field Frames and Industry Emergence: A Cultural–Political Perspective on US Recycling', *Socio-Economic Review*, 1(1), 71–104.
Luke, T. (2009), 'A Green New Deal: Why Green, How New, and What is the Deal?', *Critical Policy Studies*, 3(1), 14–28.
Lynn, S. (2012), 'Governor's Energy Office Bill Ready for Hickenlooper's Signature', *Northern Colorado Business Report*, 10 May, http://www.ncbr.com/article/20120510/NEWS/120519993, date accessed 10 May 2013.

MacNeil, R. (2013), 'Alternative Climate Policy Pathways in the US', *Climate Policy*, 13(2), 259–276.
MacNeil, R. and Paterson, M. (2012), 'Neoliberal Climate Policy: From Market Fetishism to the Developmental State', *Environmental Politics*, 21(2), 230–247.
Mahoney, J. and Thelen, K. (eds) (2010), *Explaining Institutional Change: Ambiguity, Agency and Power* (Cambridge: Cambridge University Press).
Mansfield, B. (2004), 'Neoliberalism in the Oceans: "Rationalization," Property Rights, and the Commons Question', *Geoforum*, 35, 313–326.
March, J. and Olsen, J. (1989), *Rediscovering Institutions, the Organizational Basis of Politics* (New York: The Free Press).
March, J. and Olsen, J. (1998), 'The Institutional Dynamics of International Political Orders', *International Organization*, 52(4), 943–969.
Marine Management Organisation (2012), *Marine Plan Areas*, http://www.marinemanagement.org.uk/marineplanning/areas/index.htm, date accessed 8 November 2012.
McGee, J. and Taplin, R. (2012), 'The Asia-Pacific: A Deepened Market Liberal Model for the International Climate Regime', in B. Jessup and K. Rubenstein (eds) *Environmental Discourses in Public and International Law* (Cambridge: Cambridge University Press).
McSweeney, B. (2002), 'Hofstede's 'Model of National Cultural Differences and Consequences: A Triumph of Faith – A Failure of Analysis', *Human Relations*, 55(1), 89–118.
Meadowcroft, J. (2005), 'Environmental Political Economy, Technological Transitions and the State', *New Political Economy*, 10(4), 479–498.
Meadowcroft, J. (2009), 'What About the Politics? Sustainable Development, Transition Management, and Long Term Energy Transitions', *Policy Sciences*, 42, 323–340.
Meadows, D.H. (2008), *Thinking in Systems: A Primer* (White River Junction: Chelsea Green Publishing).
Meckling, J. (2011), 'The Globalization of Carbon Trading: Transnational Business Coalitions in Climate Politics', *Global Environmental Politics*, 11(2), 26–50.
Meeus, L., Azevedo, I., Marcantonini, C., Glachant, V. and Hafner, M. (2012), 'EU 2050 Low-Carbon Energy Future: Visions and Strategies', *The Electricity Journal*, 25(5), 57–63.
Mendonça, M., Lacey, S. and Hvelplund, F. (2009), 'Stability, Participation and Transparency in Renewable Energy Policy: Lessons from Denmark and the United States', *Policy and Society*, 27, 379–398.
Meritet, S. (2007), 'French Perspectives in the Emerging European Union Energy Policy', *Energy Policy*, 35, 4767–4771.
Metcalfe, J.S. (2007) 'Innovation Systems, Innovation Policy and Restless Capitalism', in F. Malerba and S. Brussoni (eds) *Perspectives on Innovation* (Cambridge: Cambridge University Press).
Metcalfe, J.S. and Ramlogan, R. (2006), 'Restless Capitalism: A Complexity Perspective on Modern Capitalist Economies', in E. Garnsey and J. McGlade (eds) *Complexity and Evolution* (Cheltenham: Edward Elgar).
Metcalfe, S. (1995), 'The Economic Foundations of Technology Policy: Equilibrium and Evolutionary Perspectives', in P. Stoneman (ed.) *Handbook of*

the Economics of Innovation and Technological Change (Cambridge: Blackwell Publishers).

Mettler, S. (2011), *The Submerged State: How Invisible Government Policies Undermine American Democracy* (Chicago: The University of Chicago Press).

Meyer, A.D., Brooks, G.R. and Goes, J.B. (1990), 'Environmental Jolts and Industry Revolutions: Organizational Responses to Discontinuous Change', *Strategic Management Journal*, 11(Special Issue), 93–110.

Mikler, J. (2009), *Greening the Car Industry: Varieties of Capitalism and Climate Change* (Cheltenham: Edward Elgar).

Mikler, J. (2010), 'Apocalypse Now or Business as Usual? Reducing the Carbon Emissions of the Global Car Industry', *Cambridge Journal of Regions, Economy and Society*, 3(3), 407–426.

Mikler, J. (2011), 'Plus ca Change? A Varieties of Capitalism Approach to Concern for the Environment', *Global Society*, 25(3), 331–352.

Mikler, J. (2012), 'The Illusion of the Power of Markets', *Journal of Australian Political Economy*, 68, 41–61.

Mikler, J. and Harrison, N. (2012), 'Varieties of Capitalism and Technological Innovation for Climate Change Mitigation', *New Political Economy*, 17(2), 179–208.

Miles, M.B. and Huberman, A.M. (1994), *An Expanded Sourcebook: Qualitative Data Analysis* (Thousand Oaks: Sage Publications).

Milford, L. and Barker, T. (2008), *Climate Choreography: How Distributed and Open Innovation Could Accelerate Technology Development and Deployment* (Washington, DC: Clean Energy Group), http://www.sefalliance.org/fileadmin/media/sefalliance/docs/Resources/CEG_-_Climate_Choreography_7.01.08_01.pdf, date accessed 5 June 2012.

Miller, D. and Chen, M.J. (1994), 'Sources and Consequences of Competitive Inertia: A Study of the US Airline Industry', *Administrative Science Quarterly*, 39(1), 1–23.

Mitchell, C. (2008), *The Political Economy of Sustainable Energy* (Houndmills, Basingstoke: Palgrave Macmillan).

Mol, A. and Sonnenfeld, D. (2000), *Ecological Modernisation around the World: Perspectives and Critical Debates* (London: Frank Cass).

Mol, A., Liefferink, D. and Lauber, V. (eds) (2000), *The Voluntary Approach to Environmental Policy: Joint Environmental Approach to Environmental Policy-making in Europe* (New York, Oxford University Press).

Mol, A.P.J. and Sonnenfeld, D. (eds) (2000), *Ecological Modernization Around the World: Perspectives and Critical Debates* (London: Frank Cass).

Mol, A.P.J. and Sonnenfeld, D.A. (2000b), 'Ecological Modernization Around the World: An Introduction', *Environmental Politics*, 9(1), 3–14.

Molina, O. and Rhodes, M. (2007), 'The Political Economy of Adjustment in Mixed Market Economies: A Study of Spain and Italy', in B. Hancke, M. Rhodes and M. Thatcher (eds) *Beyond Varieties of Capitalism: Contradictions, Complementarities and Change in the European Economy* (Oxford: Oxford University Press).

Moody, K. (2010), 'The Direction of Union Mergers in the United States: The Emergence of Conglomerate Unionism', *British Journal of Industrial Relations*, 47(4), 676–700.

Moore, C. and Miller, A. (1994), *Green Gold: Japan, Germany, the United States, and the Race for Environmental Technology* (Boston: Beacon Press).

Morck, R. and Yeung, B. (2009), 'Never Waste a Good Crisis: An Historical Perspective on Comparative Corporate Governance', *Annual Review of Financial Economics*, 1, 145–179.

Morgan, G. and Kristensen, P.H. (2006), 'The Contested Space of Multinationals: Varieties of Institutionalism, Varieties of Capitalism', *Human Relations*, 59(1), 1467–1490.

Morgan, G.M. and Keith, D.W. (1995), 'Subjective Judgments by Climate Experts', *Environmental Science and Technology*, 29(10), 468–476.

Moser, S.C. and Boykoff, M.T. (eds) (2013), *Successful Adaptation to Climate Change: Linking Science and Policy in a Rapidly Changing World* (Oxon: Routledge).

Mowery, D. and Rosenberg, N. (1979), 'The Influence of Market Demand Upon Innovation: A Critical Review of Some Recent Empirical Studies', *Research Policy*, 8(2), 102–153.

Moylan, J. (2013), 'Colorado House Passes Controversial Renewable Energy Mandate Bill', *Steamboat Today*, 30 April, http://www.steamboattoday.com/news/2013/apr/30/colorado-house-passes-controversial-renewable-ener/, date accessed 30 April 2013.

Murillo-Luna, J.L., Graces-Ayerbe, C. and Rivera-Torres, P. (2008), 'Why do Patterns of Environmental Response Differ? A Stakeholders' Pressure Approach', *Strategic Management Journal*, 29(11), 1225–1240.

Murray, J. (2012), *Electricity Market Reform White Paper: 10 Questions that must be Answered*, http://www.businessgreen.com/bg/analysis/2092920/electricity-market-reform-white-paper-questions-answered, date accessed 10 August 2012.

Musial, W. and Butterfield, S. (2006), 'Energy from Offshore Wind', Conference Paper NREL/CP-500-39450, *Offshore Technology Conference*, Houston, Texas, May 2006.

National Audit Office (2010), *Government Funding for Developing Renewable Energy Technologies* (London: National Audit Office).

National Republican Congressional Committee (2012), 'Green is the Color of Dems' Wasteful Spending', 16 May, http://www.nrcc.org/2012/05/16/green-is-the-color-of-dems-wasteful-spending/, date accessed 10 July 2013.

National Research Council (2010), *Adapting to the Impacts of Climate Change America's Climate Choices* (Washington, DC: National Academies Press).

Negoita, M. (2010), 'To Hide or Not to Hide? The Advanced Technology Program and the Future of US Civilian Technology Policy', in F. Block and M. Keller (eds) *State of Innovation: The US Government's Role in Technology Development* (Boulder: Paradigm).

Nelson, R. (2003), 'On the Complexities and Limits of Market Organization', in J.S. Metcalfe and A. Warde (eds) *Market Relations and the Competitive Process* (Manchester: Manchester University Press).

Nelson, R. and Nelson, K. (2002), 'Technology, Institutions, and Innovation Systems', *Research Policy*, 31, 265–272.

Nelson, R.R. and Winter, S.G. (1982), *An Evolutionary Theory of Economic Change* (Cambridge: Harvard University Press).

Newell, P. and Paterson, M. (2010), *Climate Capitalism: Global Warming and the Transformation of the Global Economy* (Cambridge: Cambridge University Press).

Newell, R.G. (2008), *A US Strategy for Climate Change Mitigation*, 49 (Washington, DC: The Brookings Institution).

Noble, G. and Jones, R. (2006), 'The Role of Boundary-spanning Managers in the Establishment of Public-Private Partnerships', *Public Administration*, 84(4), 891–917.

North, D. (1991), 'Institutions', *Journal of Economic Perspectives*, 5(1), 97–112.

NREL (2012), 'Overview', http://www.nrel.gov/overview/.

O'Connor, J. (1973), *The Fiscal Crisis of the State* (Palgrave Macmillan).

O'Riain, S. (2004), *The Politics of High Tech Growth: Developmental Network States in the Global Economy* (Cambridge: Cambridge University Press).

Oakeshott, M. (2006), *Lectures in the History of Political Thought* (Exeter and Charlottesville: Imprint Academic).

Obach, B.K. (2004) *Labor and the Environmental Movement: The Quest for Common Ground* (Cambridge: The MIT Press).

OECD (1996), *Environmental Performance Reviews: United States* (Paris: OECD).

OECD (1997), *National Innovation Systems* (Paris: OECD), http://www.oecd.org/dataoecd/35/56/2101733.pdf, date accessed 2 May 2012.

OECD (2007), 'Transport', *OECD Environmental Data Compendium 2006/07* (Paris: OECD), http://www.oecd.org/dataoecd/60/38/38106409.pdf, date accessed 10 January 2012.

OECD (2009), *The Economics of Climate Change Mitigation* (Paris: OECD).

OECD (2011), *Fostering Innovation for Green Growth*, OECD Green Growth Studies, OECD Publishing, http://dx.doi.org/10.1787/9789264119925-en, date accessed 15 February 2011.

OECD (2012), *OECD Environmental Outlook to 2050* (Paris: OECD).

Official Journal of the European Union (2009), *Directive 2009/28/EC of the European Parliament and of the Council of 23 April 2009 on the Promotion of the use of Energy from Renewable Sources*, OJEU L 140/16-62.

Olsen, J. (2009), 'Change and Continuity: An Institutional Approach to Institutions of Democratic Government', *European Political Science Review*, 1(1), 3–32.

Olson, M. (1965), *The Logic of Collective Action: Public Goods and the Theory of Groups* (Cambridge: Harvard University Press).

Ostrom, E. (1990), *Governing the Commons: The Evolution of Institutions for Collective Action* (Cambridge: Cambridge University Press).

Pachauri, R.K., Reisinger, A. and Core Writing Team (2008), *2007: Climate Change 2007: Synthesis Report*. Contribution of Working Groups I, II and III to the Fourth Assessment Report of the Intergovernmental Panel on Climate Change (Geneva: IPCC).

Palan, R., Abbott, J. and Dean, P. (2000), *State Strategies in the Global Political Economy* (New York: Continuum Publishing).

Paterson, M. (2000), 'Car Culture and Global Environmental Politics', *Review of International Studies*, 26(2), 253–270.

Patterson, S. (2012), *Dark Poos: High-speed Traders, AI Bandits, and the Threat to the Global Financial System* (New York: Crown Business).

Pauly, L.W. and Reich, S. (1997), 'National Structures and Multinational Corporate Behavior: Enduring Differences in the Age of Globalisation', *International Organization*, 51(1), 1–30.

Pavitt, K. (1980), *Technical Innovation and British Economic Performance* (London: Macmillan).

Pearse, G. (2012), *Greenwash: Big Brands and Carbon Scams* (Collingwood: Black Inc.).

Peck, J. and Zhang, J. (2013), 'A Variety of Capitalism...with Chinese Characteristics?', *Journal of Economic Geography*, 13(3), 357–396.

Pelling, M. (2011), *Adaptation to Climate Change: From Resilience to Transformation* (London and New York: Routledge).

Pellow, D., Schnaiberg, A. and Weinberg, A.S. (2000), 'Putting Ecological Modernization to the Test: Accounting for Recycling's Promises and Performance', *Environmental Politics*, 9(1), 109–137.

Peng, M.W. (2003), 'Institutional Transitions and Strategic Choices', *Academy of Management Review*, 28(2), 275–296.

Perlmutter, H. (1969), 'The Tortuous Evolution of the Multinational Corporation', *Columbia Journal of World Business*, 4(1), 9–18.

Perry, T. (2010), 'Ampulse Corporation: A Case Study on Technology Transfer in the US Department of Energy Laboratories', National Renewable Energy Laboratory Technical Report.

Petrauskas, H. and Shiller, J. (1998), *Climate Change Transportation Policy* (Paris: FISITA World Automotive Congress).

Pielke Jr, R. (2010), *The Climate Fix* (New York: Basic Books).

Plant, R. (2010), *The Neo-liberal State* (Oxford: Oxford University Press).

Platzer, G. (2012), *US Wind Turbine Manufacturing: Federal Support for and Emerging Industry* (Washington, DC: Congressional Research Service).

Polanyi, K. (2001[1957]), *The Great Transformation: The Political and Economic Origins of Our Time* (Boston: Beacon Press).

Pollack, E. (2012), *Counting Up to Green: Assessing the Green Economy and its Implications for Growth and Equity* (Washington, DC: Economic Policy Institute).

Pollin, R., Garrett-Peltier, H., Heintz, J. and Scharber, H. (2008), *Green Recovery: A Program to Create Good Jobs and Start Building a Low Carbon Economy* (Amherst, MA: Political Economy Research Institute).

Porter, M. and van der Linde, C. (1995a), 'Green and Competitive: Ending the Stalemate', *Harvard Business Review*, 73(5), 120–134.

Porter, M. and van der Linde, C. (1995b), 'Towards a New Conception of the Environment-Competitiveness Relationship', *Journal of Economic Perspectives*, 9(4), 97–118.

Powell, W.W. (1991), 'Expanding the Scope of Institutional Analysis', in W.W. Powell and P. DiMaggio (eds) *The New Institutionalism in Organizational Analysis* (Chicago: University of Chicago Press).

Prévot, H. (2007), *Trop de Pétrole: Énergie Fossile et Réchauffement Climatique* (Paris: Seuille).

Pryor, F.L. (2002), *The Future of US Capitalism* (New York: Cambridge University Press).

PWC Advisory Services (2012), *Too Late for Two Degrees? Low Carbon Economy Index 2012* (London: PWC).

Quinn, J.B. (2000), 'Outsourcing Innovation: The New Engine of Growth', *Sloan Management Review*, 41(4), 13–27.
Rabe, B.G. (2003), *Greenhouse and Statehouse: The Evolving State Government Role in Climate Change* (The Pew Center on Global Climate Change).
Rabe, B.G. (2004), *Statehouse and Greenhouse: The Emerging Politics of American Climate Change Policy* (Washington, DC: Brookings Institution Press).
Rabe, B.G. (2008), 'States on Steroids: The Intergovernmental Odyssey of American Climate Policy', *Review of Policy Research*, 25(2), 105–128.
Rabe, B.G. (2010), *Greenhouse Governance: Addressing Climate Change in America* (Washington, DC: Brookings Institution Press).
Ranganathan, J., Corbier, L. and Bhatia, P. (2004), *The Greenhouse Gas Protocol: A Corporate Accounting and Reporting Standard*, revised edition (Geneva and Washington: WBCSD and WRI).
Raeburn, P. (1997), 'Global Warming: Is There Still Room for Doubt?', *Businessweek*, 3 November, 158.
Rai, V., Victor, D. and Thurber, M. (2011), 'Carbon Capture and Storage at Scale: Lessons from the Growth of Analogous Energy Technologies', Unpublished paper.
Rao, H. and Sivakumar, K. (1999), 'Institutional Sources of Boundary-spanning Structures: The Establishment of Investor Relations Departments in the Fortune 500 Industrials', *Organization Science*, 10(1), 27–42.
Rascalli, A. (2011), 'Hickenlooper's New Look COGCC', *Colorado Energy News*, 2 August, http://coloradoenergynews.com/2011/08/hickenloopers-new-look-cogcc/, date accessed 2 August 2011.
Rathzell, N. and Uzzell, D. (eds) (2012) *Trade Unions in the Green Economy: Working for the Environment* (London: Routledge/Earthscan).
Redding, G. (2005), 'The Thick Description and Comparison of Societal Systems of Capitalism', *Journal of International Business Studies*, 36, 123–155.
Renner, M. (1991), *Jobs in a Sustainable Economy*, Worldwatch Paper 104 (Washington, DC: Worldwatch Institute).
Renner, M., Sweeney, S. and Kubit, J. (2008), *Green Jobs: Working for People and the Environment* (Washington, DC: Worldwatch Institute).
Rhode Island Economic Development Corporation (2010), *A Roadmap for Advancing the Green Economy in Rhode Island*, 9 February, Version 1.0 – A Working Draft, http://publications.riedc.com/qgcju.pdf, date accessed 10 July 2013.
Ricadela, A. (2009), 'Who will be the Green VC Giant?', *Bloomberg Businessweek*, 22 October, http://www.businessweek.com/magazine/content/09_44/b4153054900653.htm, date accessed 1 April 2010.
Richter, A., West, M.A., Dick, R.V. and Dawson, J.F. (2006), 'Boundary Spanners' Identification, Intergroup Contact, and Effective Intergroup Relations', *Academy of Management Journal*, 49(6), 1252–1269.
Ritter, B. (2006), 'Nov 7 Election Night Acceptance Speech', http://ritterforgovernor.typepad.com/, date accessed 10 July 2013.
Ritter, B. Jr. (2010). 'Bill Ritter: Advancing Colorado's New Energy Economy', *Denverpost.com*, 11 March, http://www.denverpost.com/opinion/ci_14650410, date accessed 10 April 2011.
Romer, P. (1986), 'Increasing Returns and Long Run Growth', *Journal of Political Economy*, 94(5), 1002–1035.

Rosenberg, N. (1982), *Inside the Black Box: Technology and Economics* (Cambridge: Cambridge University Press).
Rosenberg, N. (1994), 'How the Developed Countries Became Rich', *Daedalus*, 123(4), 127–140.
Rosenberg, S., Vedlitz, A., Cowman, D.F. and Zahran, S. (2010), 'Climate Change: A Profile of US Climate Scientists' Perspectives', *Climatic Change*, 101(3–4), 311–329.
Rothenberg, S. (2007), 'Environmental Managers as Institutional Entrepreneurs: Environmental Managers as Institutional Entrepreneurs: The Influence of Institutional and Technical Pressures on Waste Management', *Journal of Business Research*, 60(7), 749–757.
Rothenberg, S. and Maxwell, J. (1995), *Industrial Response to the Banning of CFCs: Mapping the Paths of Technical Change*, Working Paper (Cambridge, Mass.: MIT Sloan School of Management).
Rugman, A.M. and Verbeke, A. (1998a), 'Corporate Strategies and Environmental Regulations: An Organizing Framework', *Strategic Management Journal*, 19, 363–375.
Rugman, A.M. and Verbeke, A. (1998b), 'Corporate Strategy and International Environmental Policy', *Journal of International Business Studies*, 29(4), 819–833.
Rugman, A.M. and Verbeke, A. (2004), 'A Perspective on Regional and Global Strategies of Multinational Enterprises', *Journal of International Business Studies*, 35, 3–18.
Rugman, A.M. and Oh, C.H. (2010), 'Does the Regional Nature of Multinationals Affect the Multinationality and Performance Relationship?', *International Business Review*, 19, 479–488.
Rui, H. and Yip, G. (2008), 'Foreign Acquisitions by Chinese Firms: A Strategic Intent Perspective', *Journal of World Business*, 43(2), 213–226.
Rupert, M. (1997), 'Contesting Hegemony: Americanism and Far Right Ideologies of Globalization', in K. Burch (ed.) *Constituting International Political Economy* (Boulder: Lynne Rienner).
Satter, R. (2010), 'Climategate Scientists Cleared: Questions on Climategate', *Toronto Star*, 8 July, http://web.ebscohost.com/ehost/detail?vid=5andhid=104andsid=f0ea1a67-af26-472e-bfe3-2bd74cde3240%40sessionmgr114andbdata=JnNpdGU9ZWhvc3QtbGl2ZQ%3d%3d#db=nfhandAN=6FP3981238333/, EBSCO Database, date accessed 19 October 2010.
Savage, L. and Soron, D. (2010), 'Organized Labor, Nuclear Power, and Environmental Justice: A Comparative Analysis of the Canadian and US Labor Movements', *Labor Studies Journal*, 36(1), 37–57.
Scheinberg, A. (2003), 'The Proof of the Pudding: Urban Recycling in North America as a Process of Ecological Modernization', *Environmental Politics*, 58(4), 49–75.
Schlosberg, D. and Rinfret, S. (2008), 'Ecological Modernisation, American Style', *Environmental Politics*, 17(2), 254–275.
Schmidt, V. (2002), *The Futures of European Capitalism* (Oxford: Oxford University Press).
Schmookler, J. (1966), *Inventions and Economic Growth* (Cambridge: Harvard University Press).
Schonfield, A. (1965), *Modern Capitalism: The Changing Balance of Public and Private Power* (Oxford: Oxford University Press).

Schoon, N. (2012), 'Making Britain the Adaptation Nation', *ENDS (Environmental Data Services)*, 445, 32–35.

Schrank, A. (2011), 'Green Capitalists in a Purple State', in F. Block and M.R. Kelller (eds) *State of Innovation: The US Government's Role in Technology Development* (Boulder, CO: Paradigm Publishers), pp. 96–108.

Schrank, A. and Whitford, J. (2009), 'Industrial Policy in the United States: A Neo-Polanyian Interpretation', *Politics and Society*, 37(4), 521–553.

Schreurs, M.A. and Tiberghien, Y. (2007), 'Multi-Level Reinforcement: Explaining European Union Leadership in Climate Change Mitigation', *Global Environmental Politics*, 7(4), 19–46.

Schumpeter, J. (1961[1942]), *Capitalism, Socialism and Democracy* (London: George Allen and Unwin).

Scott, S. (1993), 'Cultural Influences on National Rates of Innovation', *Journal of Business Venturing*, 8(1), 59–73.

Scott, W.R. and Meyer, J.W. (eds) (1994), *Institutions and Organizations: Toward a Theoretical Synthesis* (Thousand Oaks: Sage Publications).

Scottish Executive Development Department (2007), *Scottish Planning Policy SPP 6: Renewable Energy* (Edinburgh: Scottish Executive Development Department).

SEANERGY (2011a), *Maritime Spatial Planning (MSP) for Offshore Renewables* (Brussels: European Wind Energy Association).

SEANERGY (2011b), *Comparative Analysis of Maritime Spatial Planning (MSP) Regimes, Barriers and Obstacles, Good Practices and National Policy Recommendations* (Brussels: European Wind Energy Association).

Seo, M. and Creed, W.E.D. (2002), 'Institutional Contradictions, Praxis, and Institutional Change as a Dialectical Perspective', *Academy of Management Review*, 27(2), 222–247.

Sewell, W. (2008), 'The Temporalities of Capitalism', *Socio-Economic Review*, 6(3), 517–537.

Seyfang, G. and Smith, A. (2007), 'Grassroots Innovations for Sustainable Development: Towards a New Research and Policy Agenda', *Environmental Politics*, 16(4), 584–603.

Shaiken, H. (1986), *Work Transformed: Automation and Labor in the Computer Age* (Lexington, Mass.: Lexington Books).

Shanahan, E.A. (2010), 'The Paradox of Open Space Ballot Initiatives in the American West: A New West Old West Phenomenon', *Studies in Sociology of Science*, 1(1), 22–35.

Shnayerson, M. (1996), *The Car that Could* (New York: Random House).

Silver, H. (1987), 'Is Industrial Policy Possible in the United States? The Defeat of Rhode Island's Greenhouse Compact', *Politics and Society*, 52(3), 277–289.

Silver, H. and Burton, D. (1986), 'The Politics of State-level Industrial Policy: Lessons from Rhode Island's Greenhouse Compact', *Journal of the American Planning Association*, 52(3), 277–289.

Simpson, M. (2003), *CRS Report for Congress Climate Change: Federal Research and Technology and Related Programs*, Life Sciences Resources, Science, and Industry Division (Washington, DC: The Library of Congress).

Sissine, F. (1999), *CRS Report for Congress Renewable Energy and Electricity Restructuring*. Resources, Science, and Industry Division Congressional Research Service. Washington, DC: The Library of Congress.

Skodvin, T. and Andresen, S. (2009), 'An Agenda for Change in US Climate Policies: Presidential Ambitions and Congressional Powers', *International Environmental Agreements*, 9(4), 263–280.

Small, K. and Van Dender, V. (2008), 'Long Run Trends in Transport Demand, Fuel Price Elasticities and Implications of the Oil Outlook for Transport Policy', *Oil Dependence: Is Transport Running Out of Affordable Fuel?*, Round Table 139 of the Transport Research Centre (Paris: OECD and International Transport Forum), http://www.internationaltransport forum.org/jtrc/discussionpapers/DiscussionPaper16.pdf, date accessed 23 January 2012.

Smithwood, B. and Hodum, R. (2013), *Power Factor: Institutional Investors' Policy Priorities Can Bring Energy Efficiency to Scale* (Boston: CERES).

Solow, R.M. (1957), 'Technical Progress and Productivity Change', *Review of Economics and Statistics*, 39, 312–320.

Sorge, M. and McElroy, J. (1997), 'Ford: Grappling with Global Warming', *Automotive Industries*, 177(11), 50–51.

Sorrell, S. (2009), 'Jevon's Paradox Revisited: The Evidence for Backfire from Improved Energy Efficiency', *Energy Policy*, 37(4), 1456–1469.

Sousa, D.J. and McGrory Klyza, C. (2007), 'New Directions in Environmental Policy Making: An Emerging Collaborative Regime or Reinventing Interest Group Liberalism?', *Natural Resources Journal*, 47(2), 377–444.

Sprout, H. and Sprout, M. (1971), *Toward a Politics of the Planet Earth* (New York: Van Nostrand Reinhold).

State of Colorado, Senate Joint Resolution 98-023 (May 4, 1998).

Stenholm, P., Acs, Z. and Wuebker, R. (2013), 'Exploring Country-Level Institutional Arrangements on the Rate and Type of Entrepreneurial Activity', *Journal of Business Venturing*, 28(1), 176–193.

Stiglitz, J.E. (2002), *Globalization and its Discontents* (New York: W.W. Norton).

Stiglitz, J.E., Sen, A. and Fitoussi, J-P. (2009), *Report by the Commission on the Measurement of Economic Performance and Social Progress* (Paris: Commission on the Measurement of Economic Performance and Social Progress).

Stipp, D. (1997), 'Science Says the Heat is On', *Fortune Magazine*, 136(11), 126–129.

Stokes, B. (2012), *US Battle to Revive Manufacturing – Part I* (YaleGlobal Online).

Story, L., Fehr, T. and Watkins, D. (2012), 'United States of Subsidies', *New York Times*, 1 December, http://www.nytimes.com/interactive/2012/12/01/us/government-incentives.html, date accessed 1 December 2012.

Streeck, W. (2010), *Re-Forming Capitalism: Institutional Change in the German Political Economy* (Oxford: Oxford University Press).

Streeck, W. and Yamamura, K. (eds) (2001), *The Origins of Nonliberal Capitalism: Germany and Japan in Comparison* (Ithaca: Cornell University Press).

Stubbs, M. (2010), *Renewable Energy Programs in the 2008 Farm Bill*. Congressional Research Service (Washington, DC: The Library of Congress).

Suchman, M.C. (1995), 'Managing Legitimacy: Strategic and Institutional Approaches', *Academy of Management Review*, 20(3), 571–610.

Sweeney, S. (2012a), 'US Trade Unions and the Challenge of "Extreme Energy": The Case of the TransCanada Keystone XL Pipeline', in N. Rathzell and D. Uzzell (eds) *Trade Unions in the Green Economy: Working for the Environment* (London: Routledge/Earthscan).

Sweeney, S. (2012b), *Resist, Reclaim, Restructure: Unions and the Struggle for Energy Democracy*. Cornell University, Global Labor Institute Discussion Paper.

Szarka, J. (2008), 'France: Towards an Alternative Energy Template', in H. Compston and I. Bailey (eds) *Turning Down the Heat: The Politics of Climate Policy in Affluent Democracies* (Houndmills, Basingstoke: Palgrave Macmillan).

Taylor, C.D. (2012), 'Governors as Economic Problem Solvers: A Research Commentary', *Economic Development Quarterly*, 26(3), 267–276.

Taylor, M.Z. (2004), 'Empirical Evidence Against Varieties of Capitalism's Theory of Technological Innovation', *International Organization*, 58(3), 601–631.

Taylor, M.Z. and Wilson, S. (2012), 'Does Culture Still Matter?: The Effects of Individualism on National Innovation Rates', *Journal of Business Venturing*, 27(2), 234–247.

Tellus Institute and Sound Resource Management (no date), *More Jobs, Less Pollution: Growing the Recycling Economy in the US*.

The Economist (1997), 'Sharing the Greenhouse', *The Economist*, 345(8038), 20.

The Economist (2007), 'Out of the Dusty Labs', *The Economist*, 1 March, http://www.economist.com/sciencetechnology/displaystory.cfm?story_id=E1_RSGJRGP, date accessed 13 October 2011.

The Economist (2010), 'Blooming: Europe's Tech Entrepreneurs', *The Economist*, 10 June, http://www.economist.com/node/16317551, date accessed 24 July 2013.

Thelen, K. (2001), 'Varieties of Labour Politics in the Developed Democracies', in P. Hall and D. Soskice (eds) *Varieties of Capitalism: The Institutional Foundations of Comparative Advantage* (Oxford: Oxford University Press).

Thompson, J.D. (1967), *Organizations in Action* (New York: McGraw Hill).

Tian, L. and Estrin, S. (2008), 'Retained State Shareholding in Chinese PLCs: Does Government Ownership Always Reduce Corporate Value?', *Journal of Comparative Economics*, 36, 74–89.

Tiberghien, Y. (2007), *Entrepreneurial States: Reforming Corporate Governance in France, Japan and Korea* (Ithaca: Cornell University Press).

Toke, D. (2011), *Ecological Modernization and Renewable Energy* (Houndmills, Basingstoke: Palgrave Macmillan).

Trist, E. (1981), *The Evolution of Socio-Technical Systems: A Conceptual Framework and an Action Research Program* (Toronto: Ontario Ministry of Labour).

Trouillet, B., Guineberteau, T., de Cacqueray, M. and Rochette, J. (2011), 'Planning the Sea: The French Experience. Contribution to Marine Spatial Planning Perspectives', *Marine Policy*, 35, 324–334.

Turbeville, W. (2013), 'Gone in 22 Seconds: Just How Frequent is High Frequency Trading?', *The American Prospect*, 11 March, http://prospect.org/article/gone-22-seconds-how-frequent-high-frequency-trading, date accessed 29 July 2013.

Tushman, M.L. and Scanlan, T.J. (1981), 'Characteristics and External Orientations of Boundary Spanning Individuals', *The Academy of Management Journal*, 24(1), 83–98.

Tylecote, A. and Visintin, F. (2008), *Corporate Governance, Finance and the Technological Advantage of Nations* (New York: Taylor and Francis).

Tylecote, A., European Commission and Directorate General for Research and Innovation (2002), *Corporate Governance, Performance Pressures and Product*

Innovation, Report Number PL 98.0221 (Sheffield: European Union TSER Programme).

UJAE (2012), *Unions for Jobs and the Environment*, http://www.ujae.org/, date accessed 30 December 2011.

UK Government (2009), *The Renewables Obligation Order 2009* (London: Stationary Office).

UK Government (2010), *The Renewables Obligation (Amended) Order 2010* (London: Stationary Office).

UNCTAD (2006), *World Investment Report 2006* (New York and Geneva: United Nations).

UNCTAD (2009), *World Investment Report 2009: Transnational Corporations, Agricultural Production and Development* (Geneva and New York: United Nations).

UNCTAD (2012a), *The World's Top 100 Non-financial TNCs, Ranked by Foreign Assets, 2011*, http://www.unctad.org/Sections/dite_dir/docs/WIR12_webtab28.xls, date accessed 7 June 2013.

UNCTAD (2012b), *The Top 50 Financial TNCs Ranked by Geographical Spread Index (GSI) 2011*, http://www.unctad.org/Sections/dite_dir/docs/WIR12_webtab30.xls, date accessed 7 June 2013.

UNCTAD (2012c), *The Top 100 Non-financial TNCs from Developing and Transition Economies, Ranked by Foreign Assets, 2010*, http://www.unctad.org/Sections/dite_dir/docs/WIR12_webtab29.xls, date accessed 7 June 2013.

UNEP (2003), *Transport*, http://www.uneptie.org/energy/act/tp/index.htm, date accessed 26 May 2003.

UNEP (2008), *Green Jobs: Towards Decent Work in a Sustainable, Low-Carbon World* (Nairobi: UNEP/ILO/IOE/ITUC).

UNFCCC (1997), *Kyoto Protocol to the United Nations Framework Convention on Climate Change*, 37 ILM 22 (1998), UN Doc. FCCC/CP/1997/7/Add.1.

UNFCCC (2011), *Compilation of Economy-Wide Emission Reduction Targets to be Implemented by Parties Included in Annex I to the Convention*, 6–16 June, http://unfccc.int/resource/docs/2011/sb/eng/inf01r01.pdf, date accessed 10 May 2012.

UNFCCC (no date b), 'GHG Total Excluding LULUCF', *Time Series – Annex 1*, http://unfccc.int/ghg_data/ghg_data_unfccc/time_series_annex_i/items/3841.php, date accessed 26 July 2013.

UNFCCC (no date), *Summary of GHG Emissions for Annex 1*, available at: http://unfccc.int/files/ghg_data/ghg_data_unfccc/ghg_profiles/application/pdf/ai_ghg_profile.pdf, date accessed 18 July 2013.

United States Environment Protection Agency (2010), *Light Duty Automotive Technology and Fuel Economy Trends: 1975 Through 2010*, http://www.epa.gov/otaq/cert/mpg/fetrends/420r10023.pdf, date accessed 19 January 2012.

US Department of Defense (2010), *Quadrennial Defense Review Report* (Washington, DC: US Department of Defense).

US Department of Energy (1995), Secretary of Energy Advisory Board. *Final Report of the Task Force on Strategic Energy Research and Development* [Annex 1: Technology Profiles] June.

US Department of State (2007), *Major Economies Meeting on Energy Security and Climate Change, Sept. 27–28*, http://2002-2009-fpc.state.gov/92743.htm, date accessed 17 June 2013.

US Department of State (2008), *Asia-Pacific Partnership on Clean Development and Climate: Booklet*, http://www.asiapacificpartnership.org/pdf/brochure/APP_Booklet_English_Aug2008.pdf, date accessed 17 June 2013.

US EPA (2013), *Global Greenhouse Gas Emission Data*, http://www.epa.gov/climatechange/ghgemissions/global.html, date accessed 17 June 2013.

US Executive Office of the President (1993), *Climate Change Action Plan*, Washington, DC, October.

US Office of Technology Assessment (1995), *Renewing Our Energy Future*, September.

US State Department (2001a), *White House Climate Change Review – Interim Report, June 11, 2001*, http://www.state.gov/documents/organization/4584.pdf, date accessed 17 June 2013.

US State Department (2001b), *Paula Dobriansky: Statement to the Resumed Session of the Sixth Conference of the Parties (COP-6) to the UN Framework Convention on Climate Change Bonn, Germany July 19, 2001*, http://2001-2009.state.gov/g/rls/rm/2001/4152.htm, date accessed 17 June 2013.

US State Department (2009), *Bilateral and Regional Partnerships*, http://2001-2009.state.gov/g/oes/climate/c22820.htm, date accessed 17 June 2013.

Usunier, J-C., Furrer, O. and Perrinjaquet, A.F. (2011), 'The Perceived Trade-off Between Corporate Social and Economic Responsibility: A Cross-National Study', *International Journal of Cross Cultural Management*, 11(3), 279–302.

Van Vuuren, D., den Elzen, M., Berk, M. and de Moor, A. (2002), 'An Evaluation of the Level of Ambition and Implications of the Bush Climate Change Initiative', *Climate Policy*, 2(4), 293–301.

Vantoch-Wood, A., de Groot, J., Connor, P., Bailey, I. and Whitehead, I. (2012), *National Policy Framework for Marine Renewable Energy within the United Kingdom*, http://www.merific.eu/documents/work-package-4-policy-issues, date accessed 26 October 2012.

Victor, D. and Yanosek, K.L. (2011), 'The Crisis in Clean Energy: Stark Realities of the Renewables Craze', *Foreign Affairs*, July/August, 24–37.

Vitols, S. (2001), 'Varieties of Corporate Governance: Comparing Germany and the UK', in P. Hall and D. Soskice (eds) *Varieties of Capitalism: The Institutional Foundations of Comparative Advantage* (Oxford: Oxford University Press).

Vonnegut, K. (1952), *Player Piano* (New York: Delacorte Press).

Voss, H. (2011), *The Determinants of Mainland Chinese Outward Foreign Direct Investments* (Cheltenham: Edward Elgar).

Wade, R. (1996), 'Globalisation and its Limits: Reports of the Death of the National Economy are Greatly Exaggerated', in S. Berger and R. Dore (eds) *National Diversity and Global Capitalism* (Ithaca: Cornell University Press).

Walsh, J., Bivens, J. and Pollack, E. (2011), *Rebuilding Green: The American Recovery and Reinvestment Act and the Green Economy* (Minneapolis and Washington, DC: BlueGreen Alliance and Economic Policy Institute).

Walsh, V. (1984), 'Invention and Innovation in the Chemical Industry: Demand-Pull or Discovery Path?', *Research Policy*, 13(5), 211–234.

Wank, D. (1998), *Commodifying Chinese Communism: Business, Trust and Politics in a South Coast City* (Cambridge: Cambridge University Press).
Warner, R. (2010), 'Ecological Modernization Theory: Towards a Critical Ecopolitics of Change?', *Environmental Politics*, 19(4), 538–556.
Watson, R.T. (1999), *Report to the 10th Session of SBSTA on the Status of the IPCC* (Geneva: IPCC).
Weale, A. (1992), *The New Politics of Pollution* (Manchester: Manchester University Press).
Wee, S-L. and Popeski, R. (2013), 'China to Invest $277 Billion to Curb Air Pollution: State Media', Reuters, 28 July, http://www.reuters.com/article/2013/07/25/us-china-pollution-idUSBRE96O01Z20130725, date accessed 28 July 2013.
Weiss, L. (2003), *States in the Global Economy: Bringing Domestic Institutions Back In* (Cambridge: Cambridge University Press).
Weiss, L. (2008), 'From Military-Industrial Complex to Innovation Procurement Complex', paper presented to Berkeley Conference on Innovation Policy, Berkeley, CA.
White House (2001), *Text of a Letter from the President to Senators Hagel, Helms, Craig, and Roberts March 13 2001*, http://georgewbushwhitehouse.archives.gov/news/releases/2001/03/20010314.html, date accessed 17 June 2013.
White House (2002), *Global Climate Change Policy Book*, http://georgewbush-whitehouse.archives.gov/news/releases/2002/02/climatechange.html, date accessed 17 June 2013.
White House (2007a), *Fact Sheet: A New International Climate Change Framework*, http://georgewbush-whitehouse.archives.gov/news/releases/2007/05/20070531-13.html, date accessed 17 June 2013.
White House (2007b), *President Bush Participates in Major Economies Meeting on Energy Security and Climate Change, George W. Bush, President Remarks at the U.S. Department of State, Washington, DC, September 28, 2007*, http://georgew-bush-whitehouse.archives.gov/news/releases/2007/09/20070928-2.html, date accessed 17 June 2013.
White House (2008a), *President Bush Discusses Climate Change*, http://georgew-bush-whitehouse.archives.gov/news/releases/2008/04/20080416-6.html, date accessed 17 June 2013.
White House (2008b), *Declaration of Leaders Meeting on Energy Security and Climate Change 9 July 2008*, http://georgewbush-whitehouse.archives.gov/news/releases/2008/07/20080709-5.html, date accessed 17 June 2013.
White House (2013), *Fact Sheet: President Obama's Climate Action Plan*, http://www.whitehouse.gov/the-press-office/2013/06/25/fact-sheet-president-obama-s-climate-action-plan, date accessed 23 July 2013.
Whitley, R. (1999), *Divergent Capitalisms: The Social Structuring and Change of Business Systems* (Oxford: Oxford University Press).
Wilkinson, M. (2007a), 'APEC Soft on Emissions', *Sydney Morning Herald*, 18 August, http://www.smh.com.au/news/environment/apec-soft-on-emissions/2007/08/17/1186857771538.html, date accessed 17 June 2013.
Wilkinson, M. (2007b), 'Kyoto is the Only Way, Hu tells Howard', *Sydney Morning Herald*, 10 September, http://www.smh.com.au/news/apec/

kyoto-is-the-only-way-hu-tells-howard/2007/09/09/1189276546324.html, date accessed 17 June 2013.

Wilks, S. (1990), 'The Embodiment of Industrial Culture in Bureaucracy and Management', in S. Clegg and S. Redding (eds) *Capitalism in Contrasting Cultures* (Berlin: Walter de Gruyter).

Wilks, S. (2013), 'The National Identity of Global Companies', in J. Mikler (ed.) *The Handbook of Global Companies* (Oxford: Wiley-Blackwell).

Williams, D.O. (2012) 'Green Groups Question Lack of Representation on Denver Olympic Exploratory Committee', *The Colorado Independent*, 3 January, http://coloradoindependent.com/109048/green-groups-question-lack-of-representation-on-olympic-exploratory-committee, date accessed 10 July 2013.

Wilson, M. (2003), 'Corporate Sustainability: What is It and Where Does It Come From?', *Ivey Business Journal*, 67(6), 1.

Winter, D. (1998), 'Cleaning Up the Auto Industry's Other Tailpipes', *Ward's Auto World*, 34(8), 36–37.

Woo-Cummings, M. (ed.) (1999), *The Developmental State* (Ithaca and London: Cornell University Press).

World Bank (2008), *Donor Nations Pledge 6.1 billion to Climate Investment Funds*, http://web.worldbank.org/WBSITE/EXTERNAL/NEWS/0,,contentMDK:21916602~pagePK:34370~piPK:34424~theSitePK:4607,00.html, date accessed 20 November 2013.

World Bank (2013a), CO_2 *Emissions (kt)*, http://data.worldbank.org/indicator/EN.ATM.CO2E.KT, date accessed 6 June 2013.

World Bank (2013b), CO_2 *Emissions (kg per PPP\$ of GDP)*, http://data.worldbank.org/indicator/EN.ATM.CO2E.PP.GD, date accessed 6 June 2013.

World Business Council for Sustainable Development (2012), *Eco-efficiency Learning Module*, http://www.wbcsd.org/Pages/EDocument/EDocumentDetails.aspx?ID=13593&NoSearchContextKey=true.

Wüstenhagen, R., Wolsink, M. and Bürer, M.J. (2007), 'Social Acceptance of Renewable Energy Innovation: An Introduction to the Concept', *Energy Policy*, 35(5), 2683–2691.

Xia, J., Boal, K. and Delios, D. (2009), 'When Experience Meets National Institutional Environmental Change: Foreign Entry Attempts of US Firms in the Central and Eastern European Region', *Strategic Management Journal*, 30(12), 1286–1309.

Yanosek, K. (2012), 'Policies for Financing the Energy Transition', *Daedelus*, 141(2), 94–104.

Yin, R. (1994), *Case Study Research: Design and Methods* (Thousand Oaks: Sage Publications).

Young, S.C. (2000), *The Emergence of Ecological Modernization: Integrating the Economy and the Environment?* (London: Routledge).

Zaffos, J. (2011) 'The New Energy Economy', *Northern Colorado Business Report*, 8 April, http://www.ncbr.com/article.asp?id=57002, date accessed 8 April 2011.

Index

Abound Solar, 89, 90
absolute emission targets, 139–41
ADEME (the Environment and Energy Management Agency), 246
Advanced Research Project Agency-Energy (ARPA-E), 256, 261
Advanced Research Project Agency-Climate Change (ARPA-CC), 256, 261, 270
Advanced Technology Vehicle Manufacturing loan program, 55
Agricultural Bank of China, 230
Air Products, 144
Alcoa, 141, 146–7, 177, 181
Alliance for Sustainable Energy, 89
Alternative Motor Fuels Amendments, 50
American Automobile Manufacturers Association (AAMA), 108, 114
American Clean Energy and Security Act of 2009, 71
American Electric Power, 132, 146
American Federation of Labor – Congress of Industrial Organizations (AFL-CIO), 175, 187
American Recovery and Reinvestment Act of 2009 (ARRA), 19, 51, 72, 78, 89, 94, 178, 181, 240, 268
Anadarko Petroleum, 139, 143
anti-regulationism, 63, 69
Apache Corporation, 141
APEC Sydney Declaration, 208, 212
Apollo Alliance, 176–7
Applied Materials, 135
Aramco, 143
Arcelor-Mittal, 177, 186
Arvizu, Dan, 83
Asia-Pacific Partnership on Clean Development and Climate (APP), 51, 205–7, 211, 212
   Policy and Implementation Committee (PIC), 207

Association of South East Asian Nations (ASEAN), 205
AT&T, 177, 181, 186
Australia, 14, 98, 127–8, 158–9, 206, 208, 262, 265
   corporate perspectives, 147–57, 158
   decarbonization rate in, 158
   market mechanisms, 127
automobile industry, 14, 98, 101, 103
   emergence of, 107–9
   field emergence, 107–9
   institutional pressure transformation through organizational boundaries, 112–23
   response to climate change, 106–12
Autoworkers, 176

Bach Composite, 86
Bain Capital, 38
Battelle Memorial Institute, 55
Bayh-Dole Act of 1980, 71
Bell Labs, 7, 30, 38, 80
Blue Energy Plan, 248
BlueGreen Alliance (BGA), 15, 165, 171, 174–82, 183, 184, 185, 268–9
boundary spanners, 14, 102, 105, 112, 114, 117, 125
broader eco-efficiency, 166, 173–4
brokering, 56–7
Building and Construction Trades (BCT) unions, 174
building efficiency, 174
Bush, George W., 51, 176, 178, 190, 193, 201, 202–5, 208–11, 213
   2001 Cabinet level review of climate change policy, 202–3
   2002 *Global Climate Change Policy Book*, 203–5
Business Roundtable, 111
business unions, 171
Byrd-Hagel Resolution, 200–1

California Global Warming Solutions Act, 133
cap-and-trade system, 71, 133, 212, 265
carbon dioxide equivalent (Co$_2$e), 128, 142, 145
Carbon Disclosure Project (CDP), 14, 98, 127, 128, 129–30, 131–47, 152, 157, 159–61, 224, 258
carbon leakage, 92, 95, 180, 187
Carbon Performance Leadership Index (CPLI), 130, 131
Carbon Sequestration Leadership Forum, 51
carbon taxes, 127, 150, 155, 156, 197–8, 265
Carter, James Earl, 49, 79
centralized policies, 77, 94
CERES, 177
Charney report, 109
Chevron, 227
chief executive officer (CEO), 28, 226
China, 16, 30, 32, 131, 190–1, 208, 215–16, 219, 267, 272–3
   Agricultural Bank of China, 230
   China National Oil, 227
   China National Offshore Oil Company (CNOOC), 221, 231, 234
   Chinese Communist Party, 272
   Industrial and Commercial Bank of China, 230
   market liberalization, 219
   MNCs, 216–17, 220–5, 227, 228, 230–3, 255, 263
   VOC *see* Varieties of Capitalism (VOC)
chlorofluorocarbons (CFCs), 110, 114, 124
Chrysler, 108, 110, 185
Churchill, Winston, 273
CITIC Group, 216
Clean Air Act, 16
Clean Air Clean Jobs Act (CACJ), 87
Clean Coal Technology Solicitation Program, 50
Clean Coal Waste to Energy Program, 50

clean development mechanism (CDM), 200
Clean Energy Manufacturing Center, 177
Climate Action Plan, 51, 212, 274
   Colorado Climate Action Plan, 85
climate change regulations
   opportunities in, 135–7
   risks, 134–5, 138
Climate Change Technology Initiative, 51
Climate Change Technology Program (CCTP), 52–3, 57
   research and networking groups, 58
climate collectivism, 17, 25, 32–3, 34, 36, 251, 256, 257, 267–8, 271, 272
climate innovation process, 9
Climate VISION, 51
Climate Wise program, 112
Clinton, Bill, 51, 55, 81, 111, 200, 201
coal industry, 63, 71, 81, 92, 172, 183, 194, 260
coal-fired plants, 84, 87, 88, 146, 162, 163, 197, 199
Coal Research, Development, Demonstration and Commercial Applications Program, 50
Cogentrix, 89
Colorado, 13–14, 268
   Colorado Oil and Gas Conservation Commission (COGCC), 84
   Colorado Renewable Energy Collaboratory (CREC), 83
   Colorado Renewable Energy Act, Amendment 37, 82
   NEE *see* New Energy Economy (NEE)
commercialization, 17, 20–1, 46, 52, 55, 57, 58–9, 60, 67–8, 70, 77, 191, 235, 237, 238, 250, 260
common barriers to novel energy technology deployment, 61
Communications Workers of America (CWA), 181, 186
comparative political economy approach, 217
competition state, 49

consumer behavior, 134, 135, 136, 266
Cooperative Research and Development Agreements (CRADAs), 71
coordinated market economies (CMEs), 16, 21–3, 24, 27, 29–30, 33–4, 35, 36, 37, 38, 186, 218–19, 221, 235, 251, 253
Copenhagen Accord (2009), 255
Corporate Average Fuel Economy (CAFE), 122, 185
Corporate Climate Change Action Plan (Conoco Phillips), 141
corporate governance, 12, 28, 29–30, 36, 129, 130, 132–3, 262, 264
corporate investment, 127–9
 ineffectiveness of climate innovations, 129–31
 innovation in United States, 131–47
corporate organizational inertia, 220
corporate responses, 103–6
corporate scientists, 14, 98, 101, 105, 106, 112–23, 124
 in context, 117–21
 as monitor and filter, 112–15
corporate social responsibility (CSR), 11, 29, 30–1, 217, 224, 264
 MNC reports, 225, 228, 229
 voluntary and formal initiatives, 29
corporate-state relations, 190, 216
creative destruction of capitalism, 5
critical mass, 56
cross-shareholdings, 38
culture, and innovation, 30
Cummins, 133, 142

debt finance, 27–8, 30, 32
de-carbonization rate, 130–1, 158
decarbonize energy production, 17, 191
decentralization, 51, 79, 120, 170, 249
Deere, 138
Defense Advanced Research Projects Agency (DARPA) model, 52
Delta Air Lines, 133, 146

demand-pull, 6, 7, 8, 9, 25
Department of Defense (DOD), 69
Department of Energy (DOE), 49, 53, 55–6, 58–9, 78, 83, 89, 93, 94, 109, 112, 256
Developmental Bureaucratic States (DBS), 70
developmental network state (DNS), 13, 40, 41–2, 45–6, 48, 61, 70, 259–60, 265, 267
 causes, 47–9
 climate/energy technologies and, 49–52
 effective mechanism, requisition for, 69
 failure in energy realm, 62–7, 68–9
diesels, 4
direct public spending, 243, 246–7
dirigiste approach, 236, 242, 270
diversification of energy, 83, 86, 90, 143–7, 266
Dow Chemical, 135, 136, 137–8, 139, 141, 144–5
Dow Jones Global Total Stock Market Index, 224
Dow Jones Sustainability Index (DJSI), 14, 127, 147
Dow Jones Sustainability World Index (DJSWI), 222
Dragon Wind, 86
Duke Energy, 146
DuPont, 114, 124, 133, 142, 144

East Asia, 21, 30, 32, 58, 70, 221, 234, 257, 267
Eastman Chemical, 140, 141
eco-efficiency, 73, 74, 165–6
 broader, 173–4
 narrow, 172–3
ecological democracy, 33
ecological modernization (EM), 11, 13, 42, 72–3, 103, 165, 166–7, 174–82, 222
 American style, prospects of, 94–5
 BlueGreen Alliance *see* BlueGreen Alliance
 extractive innovations, 87–9
 and innovation, 73–5

308 *Index*

ecological modernization (EM) – *continued*
  institutional innovation, 84–7
  and political authority, 75–7
  strong EM, 13, 74, 75, 76, 87, 166–8, 171
  sub-federal, 77–9
  undoing, 89–90
  weak EM, 74–5
  see also New Energy Economy (NEE)
economic theories, 21, 32
electric utilities, 134, 142, 146, 162
emission reduction targets, 1, 42, 129, 151, 189, 194, 197, 207, 209, 271
  US resistance to, 199–205
emission trading schemes, 16, 71, 150, 157, 195, 197, 198, 200, 213, 238, 271
Energy Act of 2005, 173
energy efficiency, 5, 22–3, 41, 49, 81, 93–4, 128, 129, 135, 137, 138, 139, 142, 144, 145, 203, 212, 240, 258
Energy Frontier Research Centers, 57
Energy Independence and Security Act (EISA) of 2007, 51
Energy Innovation Hubs, 57
energy intensity, 139–41
Energy Policy Act of 1992, 49, 50
Energy Policy Act of 2005, 51, 82
energy-based economic development (EBED), 78
Enron, 28
Entrepreneur in Residence (EIR), 55
entrepreneurship, reliance on, 198
environmental federalism, 93
environmental responsibility, 217, 222
Environmental Technology Opportunities Program (ETO), 54
Europe, 16, 21, 58, 70, 111, 122, 148, 162, 250
European Union (EU), 4, 16–17, 33, 34, 35, 38, 191, 235–9, 241, 242, 247, 251, 254, 259, 270–2, 273
  agenda-setting role, 239
  directives, 238–9, 247, 251, 251
  ETS, 133, 134, 142, 162

Exchange Trading System (ETS), 133, 142, 160
  off-shore renewable energy *see* off-shore renewable energy
Exelon, 146
Exxon, 143

facilitation activities, 60–1
failed major climate bills in the US Congress, 2003–2010, 64
Farm Bill of 2008, 51
federal government, 13, 36, 45, 46, 51, 60, 67–8, 69, 71, 73, 77, 79, 89, 91, 92–4, 95, 169, 173, 177, 260
Federal Technology Transfer Act of 1986, 71
feed-in-tariff (FiT) system, 9, 198, 245, 253
Feed-in-Tariffs with Contracts for Difference (CfDs), 244
financial crises, 34, 72, 87, 88, 89, 93–4, 143, 178, 186, 197
financial incentive policies, 86, 237, 241, 243–6, 247, 253
First Environmental Action Programme, 241
Ford, 106, 111, 136, 137, 138, 139, 266
  corporate scientists at, 112–23
  interface with climate science, 116
  response to climate change, 109–10
Fortune Global 500, 222
fossil energy, market entrenchment, 62, 71, 260
fossil fuels, 3, 5, 7, 61, 62–3, 68–9, 81, 90, 92, 95, 101, 108, 129, 145, 172, 173, 184, 237, 255, 260
fracking, 61, 71, 85, 143, 260
France, 16–17, 191, 236–7, 251–2, 253, 259
  direct public spending, 246–7
  dirigiste capitalism, 191, 219, 236, 270
  energy policy, 241–3
  financial incentive policies, 243–6
  FiT scheme, 245
  Great National Loan, 246
  planning and stakeholder consultation, 247–50

Freeport-McMoran, 139
French Great National Loan, 246

Gamesa, 181
gas industry, 84–5, 87–8, 139, 143, 268
gasoline, 4, 71, 123
General Electric (GE), 90, 136, 144, 160–1
   'Ecomagination' initiative, 139
General Motors (GM), 106, 136, 144, 185, 227, 234, 266
   corporate scientists at, 112–23
   interface with climate science, 116
   response to climate change, 109–10
Geographical Spread Index (GSI), 224
Germany, 9, 27–8, 33, 38, 86, 218, 267, 271, 272
Global Change Research and Development Program (GCRD), 54
Global Climate Change Policy Book, 203–5
Global Climate Coalition (GCC), 108
global positioning, GHG emissions and, 223
globalization, 16, 77
government interventions, 8–9, 15, 37, 60, 98, 128–9, 149–4, 156, 158, 190, 256
Governor's Energy Office (GEO), 84, 86, 89–90
Greece, 23, 35
green banks, 179
Green Jobs Act, 179
green jobs, 181–2
greenhouse gas intensity targets, 203
greenwashing, 154
Grenelle II Act in 2010, 248–9
gross domestic product (GDP), 21, 38, 69, 71, 224, 272

Hickenlooper, John, 89–90
Honeywell, 135
hydraulic fracturing *see* fracking
hypothesized national measures of climate innovation, 35

IBM, 38
incremental innovation, 26, 30, 148, 253

individualism, 31–2
Industrial and Commercial Bank of China, 230
industrial ecology (IE), 74–5, 76, 80
industrial relations, 15, 167, 169–70, 180–1, 184, 185
Industry/University Research Centers, 57
ineffectiveness, of climate innovations, 129–31
information dissemination, 60
information technology (IT), 5–6, 26, 38, 26, 167
Innovation Center of the Rockies, 83
innovations at the sub-federal level, 91
Innovative Renewable Energy Technology Transfer Program, 50
In-Q-Tel, 55
insider-outsider dichotomy, 29
institutional capacity for climate innovation, 34–8
institutional entrepreneurs, 14, 104–5, 114, 115, 119
institutional factors, 164–5
   affecting union preferences, 167–71
institutional histories, and leadership, 121–2
institutional innovations, 42, 72–3, 74, 75, 85, 87, 88, 90, 91, 95, 219
   RES as, 81–4
institutional interpreters, 104
institutional pressures, 98, 102, 104, 105, 112–23, 232
institutional theory, 103–6
institutionally collective cultures, 31
Integrated Planning and Networking program, 57
integrative management-stakeholder relationship perspectives, 215
intellectual autonomy, 32
intensity *see* energy intensity
Intergovernmental Panel on Climate Change (IPCC), 193
internal commitment, 154–6
internal company strategies, 231
internal politics of labor unions, 170–1
internal rate of return (IRR), 160
International Association of Machinists, 171

International Brotherhood of
    Electrical Workers, 174
International Clean Technology Fund
    (ICTF), 209
international climate institutions,
    205–10
international emissions trading *see*
    emissions trading schemes
International Labour Organization
    (ILO), 175, 180
International Panel on Climate
    Change (IPCC), 109, 112, 113,
    114, 121, 131
  Second Assessment Report, 111
International Paper, 177
International Trade Union
    Confederation, 175
investment *see* corporate investment
Investor Owned Utilities (IOUs), 95

Japan, 4, 21, 27–8, 33, 38, 47, 200,
    205, 271
Johnson Controls, 137
joint implementation (JI), 195, 200

Krueger International, 139
Kyoto Protocol, 1, 2, 11, 23, 81, 111,
    123, 124, 137, 175, 193, 195,
    199–202, 204, 206, 207, 209, 210,
    241, 256, 271

Laborers' union (LIUNA), 174
labour unions, 164, 165–7
  BlueGreen Alliance *see* BlueGreen
    Alliance
  preferences, institutional factors
    affecting, 167–71
  technological innovation proposals,
    171–4
Lenovo, 231
liberal capitalism, 10–12, 41, 43, 76,
    95, 129, 165, 167, 171, 175, 180,
    189–90, 225, 260, 262, 267, 268,
    269, 270, 273
liberal market economy (LME), 13,
    15, 16, 21, 22, 23, 24, 27, 28,
    29–30, 33–4, 36, 37, 38, 42, 46–7,
    95, 99, 168, 186, 218–19, 221,
    236, 242, 250, 253, 262, 265, 266

Lisbon Treaty, 34
Loan Programs Office (LPO), 55
    '1703' and '1705' programs, 55
long-term investment, 9, 27, 28, 31,
    137–9

macro-institutional factors, 167–9,
    183
Major Emitters and Energy
    Consumers (MEP) process,
    208–10, 212
  MEP Leaders Declaration, 210
Major Infrastructure Planning Unit,
    248
Make It In America plan, 179–80
management autonomy, 25, 28–31,
    36, 256, 257, 261–5
managers, 17, 23, 29–31, 32–3, 97,
    105, 109–10, 115, 117, 118–19,
    121–2, 123, 124, 125, 130, 132,
    147, 261–5, 270
manufacturing plans, 179–80
Marine and Coastal Access Act of
    2009, 248
Marine Management Organisation,
    248
market forces, 19, 45, 153, 161, 191,
    232, 233
  reliance on, 2, 198, 256, 260
  versus state regulation, 226–30
market innovation, 7, 42–3, 98, 128,
    199, 213, 234, 237, 256, 260
  role in climate innovation, 46,
    152–4, 213
  versus climate innovation, 5–10, 36
market-based flexibility mechanisms,
    200
meso-institutional factors, 167, 168,
    169–70
micro-institutional factors, 167, 168,
    170–1
Microsoft, 25
mitigation policy spectrum, 197
modern states, 49
Montreal Protocol, 108, 110
multinational corporations (MNCs),
    16, 38, 106, 134, 157, 162, 187,
    190, 215, 217–18, 232–3
  CSR and sustainability reports, 225

rationales for climate innovation, 222–32
  US and Chinese, 216–17, 263
  vintage of, 219–22

narrow eco-efficiency, 172–3
National Climate Change Technology Initiative (NCCTI), 51, 52
national climate innovation systems and US liberal capitalism, 10–12
National Climate Program Act of 1978, 49, 109
national institutional variations and climate innovation, 20–5
national institutions, 13, 20, 29, 34, 41, 189, 191, 215, 217–18, 222, 235, 236–7, 241, 246, 250, 252
national planning, 250
National Renewable Energy Laboratory (NREL), 72, 79
national renewable energy targets for 2020, EU, 240
National Research Agency, 246
National Research Council, 109
nationalism, 32
nationalistic collectivism, 32
Nationally Significant Infrastructure Projects (NSIPs), 248
neoliberalism, 63, 68–9, 193–5, 260, 263, 269
  alternative international climate institutions, 205–10
  approach to global climate governance, 210–13
  climate governance, 189, 194
  environmental governance, 195–9
  resistance to binding emission reduction targets, 199–205
networking strategies, 51–2, 56–7, 58, 68, 70, 260
New Apollo Alliance Program, 176, 180
New Energy Economy (NEE), 42, 72–4, 75, 78, 79–81, 90–6, 174, 269
  ecological modernization *see* ecological modernization (EM)
  origin, 81–4
Noble Energy, 143

nomocratic institutions, 196
non-economic factors, 20, 68, 69, 186, 267
non-market cooperative relationships, 38, 218
non-market uncertainties, 266
North American Free Trade Agreement (NAFTA), 175
NSF Engineering Research Centers, 57
nuclear energy, 93, 173, 183, 242, 252

Obama, Barack, 9, 19, 51, 131, 143, 160, 161, 176, 212, 240
OECD, 10, 18, 21, 22
Office of Economic Development and International Trade (OEDIT), 86
Office of Research and Development (ORD), 54
off-shore renewable energy, in EU, 191, 235–8, 250–4
  challenges and context, 238–41
  French *see* France
  planning and stakeholder consultation, 247–50
  UK *see* United Kingdom
offshoring, 167, 170, 183
oil industry, 143, 194
organizational fields, 103, 105, 122
organizational responses, heterogeneity in, 103, 104
OSEO, 246
outsourcing, 77, 125, 167, 183

Pall Corporation, 137
Parker Hannifin, 136–7
patents, 21–3, 59, 144
patient finance, 13, 25–8, 35, 36, 237, 243, 251, 252, 256, 257–61, 270, 272
patriotism, 31
payback period, 137, 142, 162
Pemex, 143
Pew Center on Global Climate Change, 111
  Business Environmental Leadership Council, 111
pipeline model of technological development, 48, 58

planning and stakeholder consultation, 247–50
policymaking process, 16, 29, 76, 101, 120, 125
political and social institutions, 14, 20, 235
political preferences of labor unions, 171
PPG, 136, 160
Praxair, 141, 144
President's Science Advisory Committee, 109
private equity, 13, 26, 36, 38
private sector, as targets, 53
private sector unions, in US, 15
proactive market drivers, 227
programme d'investissements d'avenir, 246
public venture capital, 55

QSR Nudist (NVivo), 107, 226

Reagan, Ronald, 58, 79, 186
regional and national institutions, 191, 235, 250
Regulation Theory, 46
regulations/legislation, 5, 8, 11, 13, 23, 29, 35, 37, 51, 63, 81, 85, 88, 98, 102, 103, 108, 115, 117, 121, 122, 128, 132, 133, 134–7, 139, 141, 142, 146, 149, 153, 158, 162, 180, 195, 213, 226–30, 233, 256, 260, 263, 264, 266, 268, 269–70, 274
  modes of, 46, 47–9
renewable energy, 9, 13, 14, 16, 22, 62, 69, 77, 81, 82, 85, 86, 92, 93, 141, 143, 169, 191, 198, 208, 268, 270, 271
  and extractive innovation, 87–9
  off-shore *see* off-shore renewable energy
  relative intensity of national policies and measures (1999–2012), 24
  RES *see* Renewable Energy Standards (RES)
Renewable Energy Advancement Awards Program, 50

Renewable Energy Collaboratory (Energy Collaboratory), 90
Renewable Energy Export Technology Program, 50
Renewable Energy Innovation Fund, 246
Renewable Energy Production Incentive Program, 50
Renewable Energy Sources Act, Germany, 9
Renewable Energy Standard (RES), 73, 85, 86, 87, 90, 92, 95, 177
  as institutional innovation, 81–4
Renewable Obligation Certificates (ROCs), 243
Renewables Obligation (RO), 243
research and development (R&D), 12, 21, 23, 25, 26–7, 33, 38, 41, 45, 48, 51, 52–5, 57, 59, 61, 68, 69, 70–1, 86–7, 90, 107, 112–13, 119–20, 130, 138, 142, 144–5, 147, 153, 119–20, 257–8, 259, 264, 272
return on investment, 141–3, 152
Ritter, Bill, 72, 81, 83–8, 95, 268, 269
Rocky Mountain Innosphere, 83
Rocky Mountain Institute (RMI), 80

Schlumberger, 141
science-push, 6, 9, 25
scope 1 and 2 greenhouse gas emission, 137–8, 162, 224
Service Employees International Union (SEIU), 176
shareholder value, 27, 38, 46, 152, 217, 218, 222, 226
shareholders, 27–31, 36, 38, 97, 130, 141, 152–3, 154, 157, 185, 191, 217, 218, 219, 221–2, 226, 227, 231, 233, 261–2, 263, 264, 265
short-term investment, 9, 26, 28, 137–9, 270
short termism, 28, 36, 258, 259
Siemens, 86
Sierra Club, 176, 187
Single European Act in 1986, 241
Sinopec, 230
Small Business Innovation Research program (SBIR), 53

Small Business Technology Transfer
    program (STTR), 53
social pressure, 117, 264
social unions, 171
societal and state forces, 72, 184,
    230–1
socio-technical transitions, 182–3
Solar Energy Research Institute (SERI),
    79
solar panels, 135, 173
Southwest Airlines, 142
Space-X, 267
split estates, 85, 96
stakeholder capitalism, 28
stakeholder consultation, 242, 247–50
standards-setting, 60
state funded demonstration projects,
    60
state-level initiatives, 72–3, 78, 80, 83,
    88, 89, 91, 93, 95, 176, 180
State-owned Assets Supervision and
    Administration Commission of
    the Chinese State Council
    (SASAC), 231
Stevenson-Wydler Technology
    Innovation Act of 1980, 70
stockmarket capitalism, 27
Strategic Climate Fund, 209
strategic international business, 215
sub-federalism, 13, 14, 42, 72–3, 90–1,
    94–5, 165, 168–9, 170, 176, 177,
    178, 179, 180, 181, 183, 186
    EM, 77–9
    influences, 92–4
    innovations, 91–2
sub-federal states, role of, 168–9, 178,
    179, 183
submerged state, 76–7, 96
subsidiarity, 17, 238

targeted resourcing, 52–3, 56
Tax and Rate Treatment of Renewable
    Energy Initiative, 50
tax incentives, 60, 82, 179
technological innovations, 2, 5, 6, 8,
    10, 11–12, 13, 15, 17, 19–20, 21,
    23, 25, 27, 30, 31, 32, 34, 36,
    41–2, 47–8, 55, 77, 97, 98, 127,
    128, 148–9, 150, 157, 158, 161,
    164, 188, 189, 191, 199, 202,
    213, 235, 241, 253, 256, 257,
    265, 267
    proposals of labour unions for,
        171–4
Technology Commercialization Fund,
    55
technology investment, subsidies on,
    198
technology markets, 60
    informational and coordination
        failures in, 210–13
    state facilitation of, 198
technology transfer, 48, 53, 57–9,
    70–1, 209, 260
telocratic institutions, 196
Thatcher, Margaret, 242
3M, 133–4, 140, 144
Toyota, 5, 111
transnationality index (TNI), 216,
    223, 224
typology of climate innovations,
    166

uncertainty acceptance, 31–2, 237,
    265–7
Unions for Jobs and the
    Environment, 172
United Automobile Workers (UAW),
    176, 185
United Continental, 142
United Kingdom (UK), 3, 16, 21, 29,
    30, 31, 33, 38, 169, 191, 236, 237,
    239, 252–4, 259, 270, 272
    direct public spending, 246–7
    energy policy, 241–3
    financial incentive policies, 243–6
    planning and stakeholder
        consultation, 247–50
United Mineworkers, 183
United Nations Conference on Trade
    and Development (UNCTAD),
    216
United Nations Environment
    Programme (UNEP), 175, 180
United Nations Framework
    Convention on Climate Change
    (UNFCCC), 1–2, 193, 194,
    199–202, 204, 208, 212

United States, 1, 15, 19, 131–47, 157–8, 159–61, 239–40
  APEC Sydney Declaration, 208
  Asia-Pacific partnership, 205–7
  capitalism, 26
  Climate Change Research Initiative, 51
  climate science in context of, 122–3
  corporate governance, 132–3
  de-carbonization rate in, 130–1, 158
  Department of Energy (DOE), 109, 112
  Department of State, 206
  diversification, 143–7
  and DNS, 47–9
  economic process, 208–10
  energy intensity and process changes, 139–41
  Environmental Protection Agency (EPA), 88, 109, 112, 121
  failed major climate bills (2003–2010), 64–5
  failed non-comprehensive climate bills (1999–2010), 66–7
  federal government, 46
  global climate governance, 210–13
  labour unions *see* labour unions
  long-term versus short-term investment, 137–9
  multinational corporations, 37
  neoliberalism *see* neoliberalism
  opportunities, 135–7
  return on investment, 141–3
  risks, 133–4
  Treasury, 209
  VOC *see* Varieties of Capitalism (VOC)
United States-Asia Environmental Partnership, 50
United States Climate Action Partnership (USCAP), 71
United Steelworkers (USW), 167, 176, 181, 183, 184, 185, 186, 187
United Technologies, 143
UPS, 177
US Steel, 136, 141

Varieties of Capitalism (VOC), 41, 45, 215–17, 233–4, 235, 251, 253–4, 257
  comparative political economy approach, 217–19
  MNCs, 219–32
venture capital (VC), 12, 13, 26, 36, 38, 186, 258, 261, 274
  private VC, 56
  public VC, 55, 59
vertical integration, 30
Vestas, 86

Waste Management (WM), 133, 146
White House, 109
wind turbines, 86, 173, 181
Windsource program, 81
Wirsol, 86
Woodward Governor, 86
World Bank, 209
World Trade Organization (WTO), 175

Xcel Energy, 81, 84, 88

Yokich, Steve, 111

Printed and bound in the United States of America